Get a Job!
Third Edition

Interview Survival Skills
for
College Students

John R. Cunningham
Baylor University

Boston Burr Ridge, IL Dubuque, IA Madison, WI New York San Francisco St. Louis
Bangkok Bogotá Caracas Lisbon London Madrid
Mexico City Milan New Delhi Seoul Singapore Sydney Taipei Toronto

Get a Job! Interview Survival Skills for College Students, Third Edition

Copyright © 2005, 2001, 1998 by The McGraw-Hill Companies, Inc. All rights reserved. Printed in the United States of America. Except as permitted under the United States Copyright Act of 1976, no part of this publication may be reproduced or distributed in any form or by any means, or stored in a data base retrieval system, without prior written permission of the publisher.

McGraw-Hill's Custom Publishing consists of products that are produced from camera-ready copy. Peer review, class testing, and accuracy are primarily the responsibility of the author(s).

4567890 QSR QSR 09876

ISBN 0-07-353631-8

Editor: Barbara Duhon
Production Editor: Susan Culbertson
Printer/Binder: Quebecor World

Contents

Preface		ii
Section I	**THE COLLEGE EXPERIENCE**	**1**
	Chapter 1—Get Involved	3
Section II	**PREPARING FOR THE JOB INTERVIEW**	**11**
	Chapter 2—Types of Interviews	9
	Chapter 3—Conducting the Job Search	18
	Chapter 4—Preparing Your Resume	32
	Chapter 5—Preparing Your Cover Letter	52
	Chapter 6—Researching Prior to Your Interview	59
	Chapter 7—Behavioral Interviewing	63
	Chapter 8—Traditional Interview Questions	71
	Chapter 9—Applicant Questions	76
	Chapter 10—Interview Attire	80
Section III	**TAKING PART IN THE JOB INTERVIEW**	**97**
	Chapter 11—The Interview Opening	99
	Chapter 12—Answering Questions	103
	Chapter 13—Nonverbal Behavior	107
	Chapter 14—The Interview Closing	111
Section IV	**FOLLOWING UP AFTER THE JOB INTERVIEW**	**115**
	Chapter 15—The Follow Up	117
	Chapter 16—The Job Offer(s)	121

PREFACE

Get a Job!" Better to hear that from me than your parents. At some point in time in your life, you're going to be faced with that reality. Unless you're planning on going through life unemployed or staying at home and caring for the children while your spouse works, you're going to have to find a job! To many students, the ability to find a job and the quality of the job they get after college determines how successful their college experience was and whether or not it was worth the money that someone (you, your parents, the bank, the government, etc.) shelled out. You want your diploma to be worth more than the paper it's written on. Needless to say, it's important that you do everything in your four, five, six… years of college to get yourself ready for the afterlife (after college).

No matter how well you did in school or how great your qualifications are, you need to know what in the heck you're doing when looking for a job. There are things you have to do, steps you need to follow. That's where this book is designed to help. It's based on two different surveys. I'm not going to bore you with the all the specific details of each one, but I'll give you the cliff notes version.

The first survey interviewed 188 employers that make their bread and butter interviewing college students just like you. That survey covered every aspect of the job search, from start to finish. The employers were also asked to critique resumes and cover letters and the examples included in this book were based on their critiques. The second survey was on behavioral interviewing (covered in chapter seven), a subject that is vital to your success in the face-to-face interview. If you're not ready for it, you're going to crash and burn in the interview. That's not a threat, it's a promise. I'm not trying to scare you; I'm just trying to make you understand the importance of being prepared.

Most people look forward to taking part in a job interview about as much as going to the dentist for a root canal. It doesn't have to be a painful experience though. The best thing to do is prepare to the best of your ability. There's no better way to do that than using this book to help you with your preparation. Who better to take advice from than the people conducting your interviews? All right, so you may not be interviewed by one of *these* 188 employers, but you're going to be interviewed by someone who thinks just like them. That's close enough.

OVERVIEW

This book is divided into four distinct sections. The first section covers how you can take full advantage of your college experience so you can stock that resume with winners. The second section teaches you about the types of interviews you're likely to face, how to conduct a job search (including finding out about available jobs), writing resumes and cover letters, researching companies and positions, questions to answer (behavioral and traditional) and ask, and appropriate interview attire.

Section three covers taking part in the face to face interview, including how to make a good first impression, the keys to answering questions, important nonverbal behaviors, and how to finish strong. The final section of the book covers the follow-up process: writing thank you letters, weighing various job offers, accepting, declining and negotiating job offers.

PEOPLE TO THANK

Don't think this is cheesy. You're going to have to do it too in your job search. I had several people that were instrumental in helping me compile the information to make this book happen. I want to express my appreciation to Sherly Klepac, Lindsey Lott, Sarah Tinsley, and Blair Browning for their help in compiling the information. I want to make sure and avoid pulling a "Hilary Swank," by forgetting to thank my significant other. So, I want to thank my wife Gena for all the love and support she gives, and her ability to put up with an absentee husband during the writing of this book. You all helped make the third edition of this book a reality. I would also like to thank Dr. Charles Stewart, my mentor from Purdue that, without his initiative, I wouldn't have developed a passion for the job search process 18 years ago!

DISCLAIMERS

A couple disclaimers I need to make. First, I refer to the employer conducting the interview throughout this book as *"she."* That doesn't mean I don't think men could do that job, too. I just didn't want to always have to write "he or she said," so I made it easier on myself and flipped a coin. Second, I write like people really talk. Any English majors reading this book better get ready to deal with dangling prepositions. I dangle 'em throughout the entire book. That's something you're just going to have to learn to deal with (whoops, there's another one)!

SECTION I

THE COLLEGE EXPERIENCE

CHAPTER ONE – GET INVOLVED

"Stock Your Resume, not your Refrigerator!"

There is a lot more to gain out of going to college than a diploma and your parent's approval. Spending all your free time studying so you get good grades will benefit you in the job search, but not as much as you might think. Your parents had it easier than you (you can tell them I said that) when they were your age. If they went to college and graduated, they probably had a good job waiting for them. For most of you, it doesn't work that way. You're going to be competing for a job against a lot of other students and, while this might be hard to believe, many of those other students will be as qualified or even more qualified than you. So, what can you do to compete?

Make yourself as well-round as possible while you attend college, which doesn't mean you need to go on a Homer Simpson diet of donuts and beer. It means involve yourself in many different things on and off campus to show employers you have a variety of skills, abilities and experience they'd be crazy not to hire. Good grades are a good thing, but without any other strengths, they're not going to get you a job. I'm not arguing that good grades aren't important, just that there are other things that are more important. What's the first word that comes to mind when you think of someone that spends all of his or her free time studying? I ask that question in class every semester and "nerd" is the most popular answer. Webster's dictionary defines a nerd as "an unstylish, unattractive, or *socially inept* person." You won't find employers standing in line to hire someone who is socially inept. Employers are more likely to interview a student with 3.0 GPA and strong work experience, extracurricular activities and volunteer work than a student with 3.5 GPA and nothing else. Here are some great ways while you're in college to round yourself into shape for the job search, even if it's several years down the road:

I. MAKE FRIENDS

You've probably heard the expressions "it's not what you know but who you know" or "connections are everything." They've never been more accurate than in today's job market. There's no better way to get a leg up on the competition than knowing the "right" people. The right people are anyone that could help you get a job. Look around at the students sitting next to you in class. They may end up being the right people. It might even be their parents that are the key contacts in your job search. Or maybe your professor delivering the lecture is the chosen one. There are so many possible options to use as connections. The key is to meet them and impress them. Get to know other students in your class and in your dorm or apartment complex. Being popular is a good thing. Talk to your professors outside of class (contrary to what you might think, they're human, too, and it might even help your grades). When it comes time to look for job, any of those people potentially could be a big help. I'll talk more about connections later, but for now, just know how important it is to make friends, not enemies.

II. GET INVOLVED IN CAMPUS ACTIVITIES

It's a great idea for you to get involved during your freshman year. For most of you college represents the first time you've ever lived away from home. That, by itself, can be a very challenging time in your life. Your classes will probably require more time and effort on your part. You'll be making new friends. Surviving as a freshman, to a large extent, is about fitting in and being accepted. You may not think you have time to get involved, but it's worth a try. Getting involved in campus activities is a great way to make new friends (i.e. future connections) and make the transition from high school to college easier. Join an organization, professional or social. When the opportunity arises, volunteer or run for an office. Employers love it when students can show leadership skills in their extracurricular activities. Fitting in as a freshman is good, but standing out is even better.

III. GET INVOLVED IN YOUR COMMUNITY

I would have a hard time believing you if you told me you were too busy (save that line for your parents) to give up a couple hours of your precious free time a week to help those less fortunate than you. There are numerous organizations that rely on volunteers. Contact your local animal shelter (my personal favorite), convalescent hospital or homeless shelter (just to name a few) to see if your "free" services would be useful. You'd be amazed at how much joy you can bring to a person or animal in need. All you have to do is care. Plus, you won't be the only person to get that warm, fuzzy feeling from the experience. So will the employer looking at your resume. Be careful not to downplay your volunteer work in an interview by making it sound like it was "required." Starting out a sentence with "I *had* to volunteer as a member of my sorority" makes it sound like you wouldn't have volunteered if you didn't *have* to. What the employer doesn't know won't hurt her.

IV. GET AN INTERSHIP

It's fourth on the list of suggestions only because you'll, hopefully, make friends, join clubs and volunteer before you get an internship, however, it's number one on employers' lists. Nothing makes you more marketable on paper (i.e. your resume) than relevant work experience. The reason behind getting an internship is it shows the employer you *already* possess many of the skills they're looking for in the ideal applicant. You should start looking for an internship no later than your sophomore year in college. That will allow you to have several before you graduate.

Besides looking good on paper, internships can help you in two other very important ways. First, they often lead to a job offer. Employers don't want to hire interns that have no chance of working for them after they graduate. They'd rather hire an intern that can work for them in a career position after he or she graduates. Essentially, that's like hiring two people with one job search. Remember that piece of advice when you start an internship. If you work hard and impress the employer, there's a good chance that a permanent job offer is on its way. Think of how much more free time (i.e. fun) you can have during your senior year if you don't have to interview for a job because an internship during college led to a job offer. You can laugh at all the poor saps in their professional business attire walking around campus hoping they're employed by the time they graduate while you're laying by the pool or out shooting a round of golf. Second, an internship might actually help you figure out that you do *not* want to work for the company after you graduate. A bad experience in an internship, while very frustrating, can still be very valuable. You're a lot better off knowing a company is NOT for you during a short term internship than finding that out in a full time job after you graduate.

There are several ways to find out about internships. First, visit your campus placement center for ideas. They should have books with lists of companies that offer student internships. Some of the more popular books include:

1. The Internship Bible (Princeton Review Series)
2. Internship Success (Marianne Ehrlich Green)
3. Scoring a Great Internship (Students helping Students series)
4. Peterson's Internships 1998: More than 40,000 Opportunities to Get an Edge In Today's Competitive Job Market (Peterson's Internships)
5. The Back Door Guide to Short Term Job Adventures (Michael Landes)
6. Best 109 Internships (Best Internships)
7. Internships 2005 (Peterson's Internships)
8. Vault Guide to Top Internships (Vault Career Library)
9. The WetFeet Insider Guide to Getting Your Ideal Internship (Wetfeet Insider Guide)

Second, check with the faculty in your major. Often times, individual departments have an internship advisor that can help you locate opportunities, and you may even be able to get college credit for the internship. It doesn't get any better than that. Third, you can turn to the Internet for a list of internships. Some of the more popular web sites include:

1. College Grad.com (www.collegegrad.com)
2. GetThatGig.com (www.getthatgig.com)
3. IMDiversity.com (www.imdiversity.com)
  - For job seekers of diverse backgrounds
4. INROADS (www.inroads.org)
  - Internships for minority students
5. Internsearch.com (internsearch.com)
6. Internshipprograms.com (internshipprograms.com)
7. Internships4You.com (www.internships4you.com)
  - Must register to search for positions
8. Monster.com (www.monster.com)
9. Rising Star Internships (www.rsinternships.com)

10. SummerJobs.com   (www.summerjobs.com)
11. WetFeet.Com Internship Search Engine
             (wetfeet.internshipprograms.com)

Finally, you can create your own opportunity for an internship. Call or write the company and express your interest in an internship, emphasizing you want to work for the experience and not the money. Organizations are often willing to hire eager, hard working students, especially if it is not going to cost them any money. Obviously a paid internship would be nice, but remember the experience itself will make you more marketable and be more profitable to you in the long run.

Whichever strategy you choose for finding out about an internship, you're going to need to write a resume (Figure 1.1) and cover letter (Figure 1.2) to help you get it. Not all internships are advertised so there's an example of a cover letter for an unadvertised internship (Figure 1.3), too.

Tips for writing resumes for internships:

1. You can make the font size larger than you would on a resume for a career position. Since you probably don't have a lot of experience yet, increasing the font size will take up more space on the page and not leave you with huge margins at the top and bottom. Try a size 16 font for your name, 14 for the headings, and 12 for everything else.

2. Include the coursework you've taken so far that is relevant to the skills and abilities required in the internship. Don't limit the courses you list to those in your major. A Public Speaking class would still be very relevant for a business major looking for a marketing internship even though it is not a "major" class. Since you probably don't have a lot of work experience yet, listing coursework can help fill up the resume.

3. Include your GPA. I know, I know, I told you it wasn't that important. However, GPA can play a big role in getting you an internship, because you usually don't have a lot of relevant work experience at this point, which is one reason for the internship in the first place.

4. While including an objective is not all that common anymore on a resume for a career position (I'll talk about this in chapter four), it's still a good idea on a resume for an internship. It will help take up valuable space that you're probably having a hard time filling otherwise. A general objective will open more doors for you than a specific objective (for examples of both, read chapter four).

Tips for writing cover letters for internships:

1. Focus on what you've accomplished in school (course projects, extracurricular activities, etc.). Go into as much depth as possible without restating information an employer would be able to read on your resume. Be confident and sell yourself!

2. Try to address your cover letter to the specific person in the company that will be reading it. You can usually find this out by looking in the advertisement (if it's an advertised internship) or calling the company and asking for the information.

3. Connections are very valuable in getting internships. Use all of the resources you have at your disposal. Specific connections will be discussed in chapter three.

V. GET GOOD GRADES

I've preached up to this point that good grades aren't the most important thing to employers, but that doesn't mean you can slack off in class. A 2.0 GPA out of 4.0 will bring you minimal success in your job search even if you have solid work experience and are very involved on campus and in the community. There are some employers that still place a lot of weight on your GPA and won't even give you a second look if it's too low. One way of helping you achieve a good GPA is picking a major that is right for you. If you're happy in your major, that happiness is usually reflected in your GPA. It makes sense that you'd get better grades in classes you enjoy. If you're unhappy with your major, go talk to a counselor and explore other options. While four years might not seem like a long time to you, it will feel like an eternity if you don't like your major.

Figure 1.1 - **Example resume for an internship**

JUSTIN M. FILLER

Current Address:
254 Evans Hall
Columbia, MO 46382
(537)352-4737
E-Mail: Bigj@mizz.edu

Permanent Address:
836 Ford Road
Buffalo, NY 38476
(837)443-2882

OBJECTIVE

A summer internship in the advertising field requiring a creative and enthusiastic person dedicated to making a positive contribution

EDUCATION

UNIVERSITY OF MISSOURI, Columbia, MO
Pursuing a Bachelor of Business Administration
Major: Advertising
Minor: Public Relations
Graduation Date: May 2008
Current GPA: 3.16/4.0 (overall); 3.36/4.0 (major)

RELEVANT COURSEWORK

- Principles of Advertising
- Business Journalism
- Theories of Persuasion
- Consumer Behavior
- Creative Design
- Psychology

WORK EXPERIENCE

QUICK-E-MART, Dayton, OH
Cashier

5/05 – 8/05
- Operated cash register and performed necessary transactions
- Checked stock and recorded inventory
- Promoted to head cashier after one month

ACTIVITIES

Treasurer – University of Missouri Advertising Club
Intramural Athletic Chairperson – Kappa Omega Fraternity
Member of Organization for Outstanding Young Americans

COMPUTER SKILLS

Hardware: IBM PC, Power Macintosh
Software: Windows XP, Word Perfect 6.1

VOLUNTEER WORK

Columbia Animal Shelter

INTERESTS

Horseback riding, reading modern literature and snow skiing

Figure 1.2 - **Example cover letter for an advertised internship**

1535 Bennington Ave.
Nashville, TN 16254
March 12, 2006

Mr. Kris Lambert
College Recruiter
Top Accounting Firm
123 Park Avenue
New York, NY 18554

Dear Mr. Lambert:

I am a first year accounting major at the University of Tennessee. I discovered your advertisement for summer interns in the UT placement center on March 11th. I would like to work for and learn from an outstanding company like Top Accounting Firm.

As of May, I will have two semesters of coursework in accounting and finance. I received "A's" in both of the accounting classes I have taken at UT. I am a very dedicated worker that believes the best way to learn is through direct hands on experience. I have been active in the Accounting Club at UT and was elected Vice President at the beginning of the semester. This position has provided me with the opportunity to demonstrate my leadership skills, which would prove invaluable as I advance within a company and take on more management responsibilities.

My skills and abilities are an ideal match with those necessary for the internship. I would like to further discuss my qualifications and how I could apply what I have learned in a summer internship with your company. I will call you next week to discuss a possible interview. Please do not hesitate to call me at (635) 463-9433 if I can provide any addition information. Thank you for your time and consideration.

Sincerely,

(Signature)

Patsy Sterling

Enclosure: Resume

Figure 1.3 - **Example cover letter for an unadvertised internship**

1452 65th Street
Portland, OR 16543
April 1, 2006

Ms. Carol Higgins
Manager
XYZ Engineering Firm
647 Revolution Way
Dallas, TX 76352

Dear Ms. Higgins:

I am a sophomore at Western Plains University interested in pursuing a career in Chemical Engineering. A summer internship with your company would provide me with the opportunity to apply the knowledge learned in my first two years of coursework at WPU. I have researched XYZ Engineering Firm and your commitment to excellence and active involvement in the community are very important criteria to me when selecting a company.

In my first two years at WPU, I excelled academically with an overall GPA of 3.6/4.0 and 3.8/4.0 in my major coursework. The School of Engineering awarded me the Bob McElroy scholarship for academic excellence in my sophomore year. I am a very dedicated, hard working person hoping to make a positive difference in your company. Recently, the Young Engineer's Club elected me President, which has given me the opportunity to strengthen my communication and leadership skills. My skills, knowledge and desire fit the company philosophy of XYZ Engineering Firm.

I would like to further discuss my qualifications and how they would be utilized in a summer internship with your company. I will contact you the week of April 8th to explore this possibility. If you would prefer, you can reach me at (213) 645-2753. Thank you for your attention and consideration in this matter.

Sincerely,

(Signature)

Michael D. Dennis

Enclosure: Resume

SECTION II

PREPARING FOR THE JOB INTERVIEW

CHAPTER TWO – TYPES OF INTERVIEWS

"Pick Your Poison!"

It doesn't matter how it happens. It may be face to face, by e-mail, by regular mail, or over the telephone. Regardless of the setting, EVERY time you talk with a member of an organization, considerate it an interview and be prepared. Even if the purpose seems insignificant (i.e. getting information from a secretary about the organization), you always want to be concerned with making the best possible impression. The person you're speaking with may not have any direct influence in your hiring decision, but he or she may have a lot of influence with the person or people who do. If a secretary reports that you treated him or her rudely when you called about the position, your chances of even getting an interview would be slim.

There are three possible outcomes for every job interview. First, you can survive that round and move on to the next one. Second, you can get the boot with the majority of other applicants (such an optimist). Finally, you can get a job offer. Most job searches you'll take part in for career positions will consist of several rounds, so don't expect an offer to come rolling your way after a first interview with a company. All an employer is concerned with at that point in time is weeding out the non-winners (i.e. losers), and you just hope you're not on that list.

There are numerous types of job interviews you may have to sweat through in order to land a job. I'll go into some detail about each one so you have a better idea of what to expect. They include: (1) Informational, (2) Job Fair, (3) Phone, (4) Campus, (5) Office visit, (6) Chow time, (7) Tag-team, (8) Group, and (9) Stress. More than one of these may occur simultaneously. For example, you might take part in a Tag-team interview during an office visit. Regardless of the type of interview you take part in, be prepared!

An informational interview is not a common practice among students; but it should be. This type of interview entails you contacting a person that's currently working in a position in you want to work in when you graduate. You can make the interview even more valuable by choosing a person that not only works in your desired position, but also for a company you want to work for when you graduate. This represents an awesome opportunity to meet someone working in your career field and learn valuable information about the position and company and make a good first impression. You may have a preconceived notion of what it would be like to work in your intended career. Based on this notion, the degree to which your expectations are met usually determines the level of satisfaction you'll derive from the job and how long you'll work there. The problem is that you're not always basing your expectations on personal experience. You may be basing them on a few months in an internship, what you've read in a magazine or the newspaper, heard from friends or saw on television. It would be a lot smarter to base your expectations on what people currently working in the position tell you. You can do this if you schedule an informational interview.

To get the most out of an informational interview, know what you want to do after you graduate. Even if you were only able to narrow it down to several different career paths, use the interview(s) to speak with one or more people from each field that interests you. That way you could use this information to potentially lead you in one direction over another. If you do have one specific job in mind and you know someone currently working in that position, call her and make an appointment to talk. Conducting the interview in person is best, although a phone interview will work, too. E-mail is an option but it should be a last resort as it lacks the personal touch. If you don't know someone currently working in the position, talk with friends, past and current employers, professors, and your campus placement center to get the name of someone you can contact. The reason why it's ideal to locate a person working for the specific company you want to work for is that any information gained from her about company policy, procedures, culture, management style, etc. would be directly relevant to you. It could also help you develop questions to ask employers in a future job interview with that company. The very least you should hope to get out of this informational interview is a contact that might be able to help you in your job search. Sometimes these informational interviews lead to an opportunity to come in and interview for an actually job, but don't go into one thinking that or you might be disappointed. Make sure you take the interview seriously by preparing thoroughly and dressing and acting professionally.

A job or career fair consists of organizations sending recruiters to meet and recruit students on campus. They provide you with information about career options as well as their company, current or future job openings, hiring trends, etc. You'll never have a better chance to mass market yourself than at a job or career fair. This is an ideal opportunity for every student, especially those that have an average resume but strong communication skills. Normally, at a job fair, you hand your resume to recruiters as you are meeting them face-to-face. This allows the first impression they make of you to be based on your personality and communication skills and not just your resume. In fact, they'll probably spend very little time actually looking at your resume while you're standing there.

Some advice to follow when preparing for a job fair includes:

1. Find out prior to the job fair which employers will attend and thoroughly research those you're interested in and those that are interviewing your major. It's a good idea to prepare possible questions in advance concerning available positions, necessary skills, possible contacts, etc. Make sure you leave with any available literature the company has to offer. If you don't, it will require more effort getting it at a later date. Due to the large number of people attending job fairs, you won't have a lot of time to meet with individual recruiters, so break out your "A" game.

2. Make sure your resume is flawless (even one error can kill your chances) and bring numerous copies. Many job fairs have copiers on site, but the quality is not as good as those you can prepare in advance. You can always take copies of your resume home with you, so I would bring at least 30 copies. You normally don't need to bring a cover letter or references to job fairs (you can just to be safe), but do bring a pen

and something to write on. Your resume should have an objective at the top that states what you're interested in doing for the company. It's best to have a more general objective so you don't close any doors unnecessarily. Without an objective a recruiter may not remember what it was you were interested in when she reviews your resume at a later date. I'm sure you'd like to think you're special enough that the recruiter will remember you, but don't bank on it.

3. Be very personable and enthusiastic. Introduce yourself and tell the recruiter why you're interested in working for her company. There are probably going to be other students waiting to speak with her so you won't have much time to talk. Try and sell yourself in the brief time you do have. Get your five-minute sales pitch ready!

4. Dress professionally. That's a no-brainer. Why take a chance of being underdressed?

5. Make sure you ask each recruiter you speak with for a business card or e-mail address for follow-up discussion/correspondence.

6. After the fair is over, read any company literature you took home with you. It's a great idea to send a thank you letter to the recruiters that you spoke with personally. If you asked for that business card like I told you to, you can make sure you spell recruiters' names correctly.

A phone interview is often the first time you actually speak directly to an employer. It's a great way for a company to weed out applicants without going to the expense of traveling to talk with them. Either party can initiate the conversation. It may last five minutes or an hour, you never know for sure. You may be contacting the employer to find out more about job opportunities, or just to make sure they received the materials you sent (i.e. resume, cover letter, etc.). The employer may contact you first in response to your application materials acquired through the mail, campus placement center, Internet home page, job fair, etc. The phone interview is a very important part of the interview process. One nice thing about a phone interview is you can be walking naked around the apartment while you talk to the employer and she isn't going to know, so professional attire isn't important (even though some experts disagree and think you sound more professional if you're dressed professionally). Since the employer can't see you, that places even more of an emphasis on what you say and how you sound. You need to make sure and sound enthusiastic, speak up and in a conversational style. Answer the employer's questions thoroughly and use good grammar. Even though it may only last a few minutes, the phone interview is usually a key determinant in whether or not you get a face-to-face interview, so don't underestimate its importance.

Some advice to follow when preparing for a job fair includes: ← Someone messed up

1. Check your outgoing message. There's no guarantee that you'll be home when the employer calls, so she may get your answering machine. Many students like to have a funny outgoing message or one that has music playing in the background.

That's fine if it's a friend calling, but I don't think an employer wants to get a dose of heavy metal or rap while listening to your message. It's safest to make it short and sweet. It's also a good idea to leave something in the outgoing message that lets the employer know she called the correct number (i.e. your first or last name). If you do get a message from an employer, I'd call back ASAP, even if it's after hours. Chances are good you'll at least be able to get the person's voicemail and be able to leave a message.

2. Let everyone you live with know an employer might call. That way they won't say "He's in the can" when she asks for your whereabouts.

3. I'd leave a copy of your resume by the phone so you can refer to it, if necessary, during the interview. You can also jot down some of your key selling points (i.e. relevant skills, personality characteristics, etc.) on a sheet of paper and have it nearby.

4. Don't eat or drink anything while you're being interviewed. It can wait. Also, no gum chewing.

5. Some experts say you should stand while being interviewed because you'll sound more professional than if you are sitting or laying down. The choice yours.

6. If an employer calls, and you absolutely, positively don't have time to talk, you can try to say something like "I'm very excited you called, but I only have about 10 minutes before I have to leave. Is that enough time or would it be better for me to call you back?" By saying this, you're communicating your interest, but at the same time, telling her a later time would be better. Whether you have the guts or not to turn down a employer's request to conduct a phone interview is up to you. Some employers will expect you to make the time to talk with him or her, but most will be okay with you calling them back, just don't forget!

7. As soon as the employer hangs up, write or type a short thank you note. You can tell the employer you enjoyed the conversation, you appreciate her interest, you look forward to meeting her in person and further discussing how you are the perfect person for the job.

For many students, the campus interview represents the first chance to make a positive on the employer face to face. In the campus interview, the employer is trying to weed out students who don't have the goods they are looking for to fill the position. A common location for a campus interview is in the campus career service center. It normally resembles a factory line as one student enters a room and, in about 30 minutes, tries to convince the employer he or she is worthy of a second interview. You shouldn't be surprised if you don't get a chance to ask questions to the employer in the first interview (but still be prepared to do so). At this point, the employer isn't all that concerned with what you want to know about the company anyway. It's still looking to meet its needs for quality future employees.

In the typical job search, after the campus interview, the employer takes the information back to the company and a selection committee (it may be a committee of one if it's a small company) makes a decision on those fortunate students that get a second interview. The second round interview usually takes place on the company premises, which is how it gets its name, the "office visit" (duh). During the course of this interview, you'll get to meet numerous members of the organization. Take this opportunity to talk with people who are currently working in the position you're interviewing for. Ask them what they like/dislike about working for the company, what a typical day is like, etc. This may help you generate a more realistic perspective of what it's really like to work there.

The office visit is when salary and benefits will most likely be discussed (I'd still be prepared to talk money in a first round interview, just in case). Wait for the company to initiate the subject of money. If you bring it up first, especially in a campus interview, it makes you look like you're motivated by money. That may be true, but it's not a good image to project. This doesn't mean you should be clueless about salary. I'd research as much as possible to find out what you should expect as a starting salary. You can base it on what people are currently making in the position, your skills, education, expertise, cost of living in the city you want to work, etc. It is very possible for the employer to ask you what you expect as a starting salary. You better have a realistic answer. You may think you're worth six figures, but you'd be the only one that thinks that. It's a good idea to present a salary range to the employer rather than a specific salary figure, because it communicates flexibility on your part.

You also may be asked to take some type of test (i.e. personality, aptitude, etc.) when you go to your office visit. The good thing is you don't have to worry about studying for it. The office visit is often the last step in the interview process. The employer will let you know what you can expect next (i.e. when you can expect to hear back about your status). It's possible to receive a job offer at the end of your office visit or it may be forthcoming a week or two later either over the phone or in the mail.

There are some important things you should be looking for during your office visit. Try and get a feel for the climate in the organization. Do people seem like they enjoy working there? Do they seem genuinely interested in you as a person? Does it seem like a place you would like to work? Don't wait until you're hired to start assessing the answers to these questions.

Here are some recommendations for preparing for an office visit:

1. Be organized in your preparation. Take the name and number of the person that scheduled your visit in case you have any problems getting to the interview.

2. Make sure you have all of your travel arrangements in order. Have a route mapped out if you're driving or check your flight schedule if you are flying. If you're staying overnight, know how to get from your hotel to the office. For lack of a better expression, do a drive by of the interview site, if possible. Make sure you allow enough time for the possibility of having to deal with traffic, getting lost, parking difficulties, etc.

3. Keep close track of any expenditures by saving all receipts from interview related expenses (i.e. hotel, food, gas, etc.). Pay any unrelated expenses yourself. The trip might be paid for in advance, but you also may be required to turn in all of your receipts for reimbursement.

4. Be prepared. Research the company and position and bring the literature with you. If nothing else, it will give you something to read on the plane or in the hotel.

5. Bring extra copies of your resume, references and a copy of your transcripts. You're much better off having something they don't ask for than not having something they do ask for. Also, bring a nice pen and something to write on (i.e. leather folder).

6. Dress professionally unless the employer tells you otherwise. If nothing is mentioned about appropriate dress in advance, dress in formal business attire. As I said with the career fair, you're much better off overdressed than underdressed.

7. Remember that every person you meet throughout the entire office visit will have some say (direct or indirect) in the final hiring decision. Act professionally.

8. Immediately after you recover from the exhaustion you suffered from the office visit, type or write thank you letters to the people you interviewed with during your visit. You can either write a mass letter to everyone (if you interviewed with a lot of people) or you can personalize several letters.

There's a good chance you will go to lunch or dinner with one or more members of the company. Ah, chow time! Believe it or not there are guidelines for these meals. Some border on ridiculous. They inform you as to the proper way to put your napkin in your lap, how to fold it properly, which side of your plate you should put your napkin on to indicate you are finished eating, how to use your knife and fork, etc. Come on, that's a little too much for me. I'm not saying it's okay to tuck your napkin in your shirt like a bib, just that you have to give people some credit for having common sense.

Here are some more practical guidelines for eating a meal with a potential employer.

1. Let the host order first and then you follow. If they recommend a food item, it's not a bad idea to give it a try. Don't compromise your convictions. No one would expect you to eat a slab of beef if you're a vegetarian.

2. If you are a vegetarian, don't call the steak they order "murder." Keep those opinions to yourself.

3. Don't order the messiest thing on the menu. Unless you brought that bib with you, stay away from the ribs, spaghetti, etc.

4. Now is not the time to stick to your diet. I'm still a bit traumatized from the time I took my senior prom date to the best steakhouse in town and she ordered a salad… and nothing else. Sure she was a cheap date, but the old man a gave me a wad of cash and I was aching to look like a big spender. The employer is going to write off the meal anyway, so don't worry about being thrifty. Of course, you don't have to get carried away and order the most expensive entrée either. If you're asked to order first, you can get a suggestion for an entrée from the employer. Remember you're going to be a little nervous which is going to shrink your appetite.

5. Ah, now it's time to discuss the all-important choice of beverage question. The server approaches your table and asks you what you'd like to drink. What do you order? Personally, I'd recommend staying away from alcohol during an interview unless the company orders it first. A glass of wine isn't going to kill you, but if it's been known to make you loopy in the past, don't order alcohol regardless of what the employer orders.

6. I made fun of all the specific rules for your napkin earlier, but that doesn't mean I think good table manners aren't important. Follow the basics. Keep your elbows off the table and your napkin in your lap. Don't talk with food in your mouth and don't slurp your drink or soup. Do what your mom (hopefully) taught you to do and you should be fine.

7. During your meal, don't get fooled into thinking that you can let your guard down just because you aren't in an office. This is still a very important part of the evaluation process. The employer may be assessing your social graces as well as your communication and interpersonal skills. Avoid saying or doing anything that could be held against you.

How would you feel if you showed up for your office visit, walked into a room and discovered you're going to be interviewed by three members of the organization at once? Or what if there were four other students with you in the room waiting to be interviewed by one recruiter? Any extra tension? You better believe it! The tag-team and group interview approaches represent opportunities for the employer to accomplish several things: (1) Save valuable time (time is money) and (2) Make more reliable comparisons of applicants and (3) See how you handle the added pressure. They save time because the alternative to a tag-team interview is a sequential interview process where one person interviews you then passes you on to the next and so on and so on. If the employer wanted four people to interview you for 30 minutes each, it would take two hours do finish. With the tag-team approach, it would only take 30 minutes.

Comparisons are generally more reliable because each interviewer is witnessing the EXACT same answer to each question. Finally, there's more pressure in both the tag-team and group interview setting. The employer wants to know you can handle stress BEFORE you start the job!

You're fortunate that the tag-team and group interview aren't all that common yet, but I've noticed an increase in both types over the past several years. You need to realize that both are a possibility, so be prepared just in case it's not your lucky day. If you're in a tag-team interview, make eye contact with all of the people when answering a question, and not just the person that asked it. If you're in a group interview, take the opportunity to demonstrate in your answers why you're a more qualified applicant than any other person in the room.

The last type of interview is called a stress interview. You might be thinking, "Isn't *every* interview a stress interview?" The correct answer would be yes, of course, but an employer specifically conducting a "stress" interview intentionally tries to make the stress level high to see if she can get you to crack under the pressure. Usually stress interviews are reserved for jobs that have an inordinate amount of stress (i.e. nuclear power plant worker, police officer, etc.) Very few students ever get to experience a stress interview, which is a good thing. Just in case though, be ready. The most important thing of all is NOT to lose your calm or you're going to lose any chance of getting the job!!! How do you know if you're in a stress interview? I'll tell you. The employer might make you wait a ridiculously long time (30 minutes or an hour) before greeting you. The greeting may be cold and aloof, with minimal eye contact. Rapport building might be non-existent. The employer may intentionally be rude or critical of you or your answers during the interview. All of this is designed to see how you handle the pressure. While it might feel great to dish it right back to the interviewer, avoid the temptation. Also, try not to take it personally. The employer was the same to the applicant before you and will be the same to the next applicant. Survival is the key!

Regardless of the type of interview, there are several recommendations. First, be prepared. Research the company and position thoroughly before you talk with an employer. Ask intelligent questions that indicate you've done your research and are very interested in the company and position. Dress professionally. Be professional, enthusiastic and sincere (all valued characteristics by an organization). Demonstrate relevant skills and abilities. Make sure you follow up with a thank you letter after each interview. Handwrite a personal letter to the person or people you interviewed with. Mention something that impressed you about the company. You can also mention something you talked about during your interview that might help them remember you specifically. If you see a picture of a dog on the employer's desk, become a dog lover for the interview and in the thank you letter.

Chances are good that no two interviewing experiences you have will be the same. This can stem from the type of interview situation you are in or individual differences in the people who are conducting the interviews. No matter how prepared you are for the different types of interviews, chances are good that something unexpected will occur. You can usually thank the employer for that. Employers differ in terms of their philosophy and the importance they place on the employment interview. It only takes one experience with an untrained interviewer to realize that, regardless of the situation, the ability of the interviewer plays a huge role in the process and outcome of the interview. The next chapter starts you on the path to preparing for the job interview.

      Hopefully, as your college career is coming to an end, you have a clear idea of what you want to do after graduation. While some students decide to put reality on hold for a couple of years while they attend graduate school, the majority still choose to find a job. Finding a job isn't as easy as it used to be because there are so many qualified applicants. As I mentioned in an earlier chapter, applicants used to be considered qualified and even stood out among other applicants if they possessed that all-important college diploma. That same college diploma today may look nice in a frame on your wall, but you can't rely on that piece of paper alone to get a job. Most of the other applicants you're competing with will have that same piece of paper, but theirs may be in a nicer frame. While the job market determines, to an extent, the number of jobs you can apply for, there are steps you can take to help out your cause and move ahead of the pack. One of the key steps is preparation. The more thoroughly you prepare for the job interview, the better your chances will be to find employment. The chapters in part II cover the steps that should be followed during the preparation process.

CHAPTER THREE – CONDUCTING THE JOB SEARCH

"Pound the Pavement!"

Before you even begin conducting the job search, you must first get to know the product you're selling better than you have ever before. If you don't know by now, YOU are the product. I'm sure you're thinking you're an expert on that topic because you've been living with yourself your entire life. Don't be fooled into thinking this, because you need to know yourself better than ever to determine if you're entering the career best suited for your skills, abilities, interests, etc. You're going to be asked questions about yourself in interviews that you probably hadn't given much, if any, thought to in the past. To determine the job you're best suited for, you have to figure out what you value in a position. Determine which of the following work values (there may be others that aren't listed) are important to you:

- Interesting work
- Job security
- Recognition for work done
- Salary and benefits
- Good working environment
- Type of supervision
- Contribution to society
- Authority

- Advancement opportunities
- Diversity of tasks (job rotation)
- Amount of creativity
- Co-workers
- Teamwork
- Flexibility
- Independence

Once you know which values are important to you, analyze different jobs to see which of them contain the values you need to be happy. Once you've accomplished this, it's time to get to know yourself even better. The best method of doing that is to conduct a self-analysis. This entails answering questions about your education, skills, work experience, goals, strengths and weaknesses, etc. The following questions (answers listed are given only to guide your thinking) provide a good focus for your self-analysis[1]:

1. What are your personality strengths and weaknesses?

   - Reliability
   - Dynamic
   - Flexibility
   - Honesty
   - Diligent

   - Motivation
   - Values, morals
   - Enthusiasm
   - Innovative
   - Cooperative

   - Open-mindedness
   - Maturity
   - Self-starter
   - Patient
   - Resourceful

---

[1] Adapted from Charles J. Stewart and William B. Cash, Jr., **INTERVIEWING PRINCIPLES**, 8th edition. Copyright © 1997 Times Mirror Higher Education Group, Inc., Dubuque, Iowa. All Rights Reserved. Reprinted by permission.

2. What are your intellectual strengths and weaknesses?

   - Intelligence
   - Analyzing
   - Planning
   - Team building
   - Quick learner
   - Creating
   - Reasoning
   - Problem solving
   - Organizing
   - Evaluating
   - Decision making
   - Implementing

3. What are your communicative strengths and weaknesses?

   - Oral communication
   - Interpersonal skills
   - Teaching
   - Written communication
   - Persuading
   - Articulate
   - Listening
   - Negotiating
   - Clear

4. What relevant skills can you offer and what specific examples can you provide?

   - Communication
   - Computer
   - People
   - Leadership
   - Problem solving
   - Time management
   - Organizational
   - Persuasive
   - Analytical

5. What have been your accomplishments and failures?

   - Academic
   - Professional
   - Extracurricular activities
   - Personal
   - Work
   - Goals

6. What are your professional strengths and weaknesses?

   - Education
   - Training
   - Experiences
   - Specific abilities
   - Talents
   - Specific knowledge

7. What are your professional interests?

   - Short range goals
   - Advancement
   - Long range goals
   - Recognition
   - Income
   - Status

8. Why did you attend _____ (your college) and how happy are you with your decision?

9. Why did you major in _____ and how happy are you with this decision?

After you've analyzed yourself and know exactly what you can offer an employer, it's time to create a list of potential companies and positions that would satisfy your needs.

The first step in the job search process is finding out about available jobs you're interested in. As a general rule, the more flexible you're the more available jobs you will find. A key to flexibility is a willingness to work in any city. People who want to live and work in a particular city will have a more difficult time finding as many available job openings as someone who is willing to live and work anywhere. One option in finding out about job openings is to send cover letters and resumes to companies you want to work for and hope they have a position available for you. Unless you enjoy rejection, this isn't a very wise strategy to try. Your chance of even getting an interview in this scenario is slim. You'll be much more successful if you target your search by applying for jobs that companies are advertising to fill or using connections to find those "unadvertised" positions. As far as timing is concerned, you should begin your job search about 6 months before you plan to start working. There are several excellent sources to utilize when looking for available job openings:

1. Networks

I talked about the benefit of having connections to help you get a job in chapter one. The process of making those connections is called networking. Roughly 80% to 85% of the job openings available at any one time are never advertised, so it's imperative for you to establish networks to find those "hidden" jobs. The first step in establishing a network is determining possible contacts. Make a list of as many people as you can think of that might make a valuable contact. Sources to utilize include:

- Professors
- Employers (past or current)
- Friends and relatives
- Alumni
- Social/Professional organization members
- Campus placement center counselors

Your list doesn't have to be limited to people you know directly (I mentioned your friend's parents in chapter one). Maybe one of the individuals listed above knows someone that you may be able to use as a contact. It's a good idea to keep a written record of all of your contacts. Take advantage of all possible connections and make sure and show your appreciation (as simple as a hand written thank you) to anyone who provides assistance. You can contact sources by mail, e-mail or telephone. Whether you call or write, introduce yourself and mention who referred you if you do not know the person you're speaking with. Identify your reason for calling/writing and try to set up a time that would be convenient to meet to discuss your situation. Try and meet with this contact in person. If that's not possible, see if you can fax her a copy of your resume (bring copies with you if you meet in person). Be prepared! Before you contact this source, make sure you know exactly what it is you're looking for in terms of a position. Think about the questions you want to ask in advance so you don't arrive at the interview without any questions in mind. If you're relying on your ability to think of appropriate questions to ask on the spot, you'll probably discover it's not as easy as you think. Be professional. Remember, this person's opinion of you may play a significant role in your chances of getting a job with the company he or she recommends. Before you leave the interview or hang up the phone, make sure you have a business card or at least the correct spelling of her last name for follow-up correspondence. It's also a good idea to let anyone who helped you secure a job know about it once you've accepted. They may not be as excited as you, but it's still the considerate thing to do. Who knows? You may need to have them help you again at some point in time in the future. You don't want to burn any bridges in the process.

2. Campus Placement Center

Most colleges have placement centers designed to help students with the job search process. This includes assisting with the different aspects of preparing for the job search (resumes, cover letters, researching companies and positions, mock interviews, etc.) They also invite companies to come to campus to interview students for job openings. Some placement centers charge a fee, but it's minimal compared to more expensive placement agencies and headhunters. I strongly recommend checking into what your placement center can do for you because you have nothing to lose. I've heard the complaint that some placement centers only help certain majors. While it may be true that some majors are helped more than others, it certainly can't hurt to give them a chance. As long as the placement center has your resume in its database, any company contacting them to do a resume search from the database can discover you. Not all companies are going to send employers to campus. Many choose to conduct the database search because it's more cost effective and they'll determine who to interview based on this search. There are strategies to follow when creating a resume for a database, which will be discussed in the next chapter. If you do decide to avoid the placement center and go out on your own, you'll probably be spending a lot more time and money, and you can certainly expect more frustration.

3. Employment Agencies

Don't confuse these with a college placement center. These agencies will help you try to find a job but it will probably cost more than a placement center. The reason I wrote "probably" is that you may be able to negotiate so that the company that hires you agrees to pay the fee. However, that's no guarantee. The fee can be up to several thousand dollars (% of your first year salary) which is why I recommended earlier that you at least try your campus placement center. There's no guarantee that you'll get a job even with the assistance of an employment agency. One benefit of an employment agency is that it will do the legwork for you. They'll match up your qualifications with available jobs and even go so far as to set up interviews with these companies. However, it's still up to you to get a job offer from the interview. It's important to realize that these agencies don't know about all of the available jobs. As I mentioned earlier, only 15% to 20% of the available jobs out there are ever advertised. An employment agency will know about more jobs than the newspaper, but there will still be a lot of "hidden" jobs. Make sure they're only sending you to job interviews you're qualified for or it could end up being a waste of your time and money.

4. Newspapers

You can use the newspaper to find out about a lot of available jobs at one time. You can use your local newspaper or you can use the local paper from the city you want to work in when you graduate. Most city newspapers are now available on the internet, which includes the classified ads and the job listings. You may not find the newspaper to be very useful (many ads will be for lower level jobs than you're seeking or temporary positions), but it never hurts to try.

5. Reference Books

These books can help you fulfill that goal you set out to achieve when you started college: getting a job. These sources list relevant information on companies and jobs. They include (this is only a partial list):

- The Adams Job Almanac
- The 100 Best Companies to Work for in America
- Career Employment Opportunities Directory
- College Placement Council Directory
- Hidden Job Market
- The National Job Bank
- The National Jobline Directory
- Occupational Outlook Handbook
- Peterson's Job Opportunities for Business and Liberal Arts Graduates
- Peterson's Job Opportunities for Engineering, Science and Computer Graduates
- Professional's Job Finder
- Where the Jobs Are
- World Chamber of Commerce Directory

6. Computers

You would probably have a hard time finding someone your age that doesn't know how to surf the internet. Now you can actually use those skills for good, instead of just to kill time. The Internet is rich with information on careers, organizations and job openings. Figure 3.1 contains a list of Internet sites to help you with finding information about careers, companies and available jobs.

Once you've located an appropriate number (which will vary depending on the individual student) of available jobs, you can begin the application process. Remember that the more jobs you apply for, the more likely you'll be to get invited to an interview. You need to apply for a workable number of jobs. Your definition of workable depends on you. Every interview you get requires hours of preparation time going over possible questions and answers, researching the company and position, thinking of good, quality questions to ask, etc. If you're lucky enough to send out 30 resumes and get 30 interviews, preparing for those interviews will become your full-time job. However, if you have a "put all of your eggs in one basket" approach and only apply for one job and come up empty, you could suffer some unnecessary discouragement.

Let's say you apply for 15 jobs and get 8 interview offers. At the same time you're waiting tables in a local restaurant 20 hours a week and going to class full time. In this case, preparing for 8 interviews is a lot. If you can handle it, more power to you. If you can't, try prioritizing your interviews. Determine the two or three jobs you would like to have the most and prepare for these very thoroughly. Then, take the middle two or three jobs and spend some time preparing for these, but not as much time as you spend on the top two or three. Finally, spend minimal time preparing for the jobs at the bottom of your list, those jobs that you would be interested in only if none of the top or middle jobs pan out. If possible, try and schedule the job interviews you're least interested in first. Use them as practice, especially if you're a rookie in the job interviewing game. You don't want to have that dream job be the first one you interview for, because you are inevitably going to experience those "first ever interview" nerves (for lack of a better term). I've jumped ahead a little bit and started talking about interviews. Remember though, before you get an interview you need to prepare a quality resume, and that topic will be discussed in Chapter four.

Figure 3.1 – **Internet Sites**

Due to the fact that the Web is updated frequently, Internet addresses can change. It is a good idea to check that you are using the most current version of the information. If you are interested in Internet Job Searching, there are several good books you can use to help you in your search:

1. *Power Searching the Internet for Your Next Job* by Thomas E. Wright
2. *Job Searching Online for Dummies* by Pam Dixon, Dummies Technology Press)
3. *Guide to Internet Job Searching 2000-01 Edition* by Margaret Dikel, Frances E. Roehm and Joyce Lain Kennedy
4. *Adams Electronic Job Search Almanac 2000* (Adams Electronic Job Search Almanac) by Thomas F. Blackett (Editor), et al.
5. *110 Best Job Search Sites on the Internet* by Katherine K. Yonge

As far as web sites are concerned, here is a list of sites that can help you with all aspects of the job search process:

1. **America's Job Bank** – (www.ajb.dni.us/)
   Job openings from all over the country are listed. Searches can be performed by keyword, use of subject menu, different classification codes. You can also conduct a geographic search.

2. **Best Jobs in the USA Today** – (www.bestjobusa.com)
   The Best Job's database allows you to search for a job, narrowed by geographic location, broad categories and key words.

3. **Career Builder** – (www.careerbuilder.com)
   This searchable database provides a number of criteria to focus on, but the geographic options are limited to California, Maryland, New Jersey, Texas, Virginia, and Washington D.C. At no charge, you can establish a profile with Career Builder and be notified of opportunities meeting your criteria.

4. **Career City** – (www.careercity.com)
   Includes newsgroup listings and corporate recruiting links.

5. **Career Magazine** – (www.careermag.com)
   Includes a job openings' database of job postings from major newsgroups, a resume bank, employer profiles, articles and news which can help you with your career search, and a career forum where you can share experiences and seek advice.

6. **Career Shop** – (www.careershop.com)
   Jobs, resumes, personal job shopper, and virtual job fair.

7. **Clearance Jobs** – (www.clearancejobs.com)
   This database is dedicated to employment opportunities requiring a security clearance.

8. **College Grad Job Hunter** – (www.collegegrad.com)
   This database focuses on entry-level positions and can be browsed or searched by key words.

9. **Computer Job Store** – (www.computerjobs.com)
   Search jobs by location and post your resume.

10. **Cool Jobs** – (www.cooljobs.com)
    This page provides links to cool places to get non-technical jobs, such as Carnival Cruise Lines, Club Med, Liz Claiborne, MTV, Walt Disney, just to name a few.

11. **DICE** – (www.dice.com)
    The premier job search website for computer professionals with thousands of high tech, permanent, contract and consulting jobs available. It searches based on key words.

12. **Ecology employment** – (www.ecoemploy.com)
    Job opportunities for careers relating to the environment such as environmental engineers, chemists, geologists, and wastewater treatment.

13. **Helmuts to Hardhats** -- (www.helmutstohardhats.org)
    Employment opportunities for transitioning military personnel: construction to professional jobs are listed at this database.

14. **Help wanted.com** – (helpwanted.com)
    Search for positions by company name or keyword. Resumes can be posted at no cost.

15. **Hot Jobs** – (www.hotjobs.com)
    Choose from many career channels at entry level or from start up careers.

16. **Internet Career Connection** – (www.iccweb.com)
    Maintains a searchable "Help Wanted" database of positions.

17. **Job Bank USA** – (www.jobbankusa.com)
    The JobBank MetaSEARCH is your gateway to Internet Employment Resources. Search the JobBank USA database, other databases, and newsgroups from one convenient location.

18. **Job Center** – (www.jobcenter.com)
    Provides access to a database of job advertisements and resumes, both of which can be searched. Resumes and job advertisements can be posted for a fee and are distributed to USENET newsgroups and to job seekers or employers with matching resumes or job ads listed at the service.

19. **Jobs.com** – (www.jobs.com)
    Build your own myjobs.com page and choose only the information you want.

20. **Jobs.net** – (www.jobs.net)
    Job opportunities by category and/or location.

21. **Job Direct** – (www.jobdirect.com)
    Focuses on entry-level positions and internships; free service for job seekers.

22. **Job Trak** – (www.jobtrak.com)
    This job-listing service partners colleges and universities throughout the country to provide access to job postings specifically for the students and alumni of the institutions. You may search for jobs and internships and post resumes (in ASCII) at no charge and easily forward your resume to an employer for any advertised position.

23. **Job Web** – (www.jobweb.org)
    Owned by the National Association of Colleges and Employers, this site provides keyword searching by job postings or employer directory. Searches can be narrowed to federal jobs or internship/co-op opportunities and national or international locations. Links are also provided to additional resources that can be used for career planning and employment information.

24. **Monster.com** – (www.monster.com)
    A database of job listings that can be searched by keywords, industry, or geographic regions. Also provided is information on internships and opportunities at colleges and universities. Resumes can be posted at no cost.

25. **Nation Job Network** – (www.nationjob.com)
    Select from a list of extensive criteria including field, geographic location, duration, education, and salary to search for a position or company that meets your needs.

26. **Recruiters Online Network** – (www.recruitersonline.com)
    "Job seekers find thousands of jobs and recruiters in your field or location." Free job search and resume posting.

27. **The Ladders** – (www.theladders.com)
    Job opportunities of $100K+ for executives are found at this database.

28. **True Careers** – (www.truecareers.com)
    Database for employment of degreed professionals.

29. **Top USA Jobs** – (www.topusajobs.com)
    Database for job opportunities by geographical location, by categories, or by narrowing the search by use of keywords.

30. **USA Jobs** – (www.usajobs.opm.gov)
    the official job site of the United States Federal Government.

31. **Wall Street Journal** – (careers.wsj.com)
    Search for jobs by location, industry or company.

32. **Yahoo! Careers** – (careers.yahoo.com)
    Search over 1,000,000 jobs nationwide. Build and edit your resume. Research salaries, companies and industries.

There are also many web sites that specialize in helping people in a particular field. Included in this list are:

1. **AIP Career Services** – (www.aip.org/careers)
   Sponsored by the American Institute of Physics, from this site you can learn about their resume search service, job centers, career workshops, and access job opportunities.

2. **American Medical Association** – (www.ama-assn.org)
   Physician recruitment advertising is accessible from the Journal of the American Medical Association. To access the AMA publications, you have to register, but there is no fee for the service.

3. **Career/Employment Information Compiled by Scholarly Societies** (www.lib.uwaterloo.ca)
   One way to find positions in a particular discipline is to look for announcements from within the professional organizations. This page provides direct access to those Scholarly Societies that have career and employment information available on their web sites. If

a particular society or association is not listed on this page, you may want to look on the Scholarly Societies Project home page for the society or organization of interest.

4. **Chronicle of Higher Education Job Announcements –** (chronicle.com/jobs)
   Browse by broad subject category or by broad geographic division for faculty and administrative positions available in colleges and universities.

5. **Entertainment Recruitment Network –** (www.showbizjobs.com)
   They say it is tough to break into show business at any level, but this surely will help. It covers film, TV, recording, multimedia, and theme parks, including internships (unpaid!).

6. **Environmental Careers Organization –** (www.eco.org/)
   A national, nonprofit, educational organization dedicated to the development of individuals' environmental careers. Provides information about career publications, environment career job fairs, as well as links to sites with career information for environmentalists.

7. **Fed World: The U.S. Government Bulletin Board –** (www.fedworld.gov/categor.htm)
   The premiere location for jobs with the federal government, including the Department of the Interior Automated Vacancy Announcement System, the Office of Personnel Management Job Announcements, and the Peace Corps.

8. **GeoWeb Marketplace –** (www.ggrweb.com)
   Positions in geoscience and technology are listed here.

9. **JobSmart – Specific Careers** (www.jobsmart.org/tools/career/spec-car.htm)
   Organized by discipline, the resources on this page mostly provide career information in specific fields. Some of the resources also provide links to finding jobs in the specific discipline.

10. **Legal Employment Listings –** (www.lawjobs.com)
    Available online daily, hundreds of legal employment listings from the National Law Journal, Law Technology Product News, and many more additional online listings.

11. **MedNet** – (www.healthjobsusa.com)
    A searchable database from <u>Best Jobs in the USA,</u> that focuses on opportunities in the healthcare profession. Searches may be narrowed be geographic location.

15. **SHS Career Resources** – (www.shsinc.com/)
    This web site focuses on careers in the Telecommunications or Medical Communications industry, including sales positions.

16. **Sports Link – Employment Opportunities**
    (www.sportlink.com)
    SportLink provides a wide variety of positions available in the sports and recreation industry, listed by dates and job title. This site is sponsored by the Sporting Goods Manufacturers Association.

17. **TechCareers** – (www.techweb.com/careers/careers.html)
    This site focuses on careers in high technology and provides access to a searchable database of over 10,000 job announcements and in conjunction with <u>E-Span,</u> brings you TechHunter, a free service that: maintains a personal profile of your job preferences and qualifications; lets you know which jobs match your profile; allows you to apply immediately and on-line for many jobs; e-mails job opportunities to you; tells you what is new in your profile area every time you visit.

18. **Technical-Web Multimedia Job Hotline** (www.nwu.org)
    This is a nation-wide project of the National Writers Union and is run by, and for, the writers who use it. The names and phone numbers of the companies, employers, and managers who list jobs are only given out to members of the NWU. Writers contact and deal with employers directly. Based on an honor system, writers who find a job through the Hotline listing pay a finders-fee of 10% of their first four months' income. The announcements are simply listed and cannot be searched.

19. **TV Jobs** – (www.tvjobs.com/)
    Advertised positions are organized in four categories (television, radio, cable, education) and displayed in tables which provide a brief overview of the position and link to a more complete description. There is also a "Job Bank" from which you can view employment pages of TV stations across the country. Resumes can be posted at this site for a fee.

20. **University of Minnesota's College of Education Job Search Bulletin Board** – (sps.coled.umn.edu/students/career/index.html)
    This "bulletin board" pulls together hundreds of brief job postings focused on the K-12 educational field. Categories include administration, media specialty, psychology, and counseling; early childhood development and elementary education; higher education; and secondary, special, vo-tech education, and physical education. The available positions can be browsed by specialty.

21. **Veterinary Jobs on the Net** – (www.metronet.lib.mn.us/lib/pathfind.htm)
    This page is dedicated to job postings for veterinarians who are looking for a job, clinics that are looking for veterinarians and locum postings. For a fee, resumes and CV's can also be posted at this site.

If you're interested in working in a particular city or state or foreign country, there are web sites that can help you with a geographical search, including:

1. **Best Bets for Working Abroad** – (www.cie.uci.edu/~cie/iop/work.html)
    Provided by the University of California at Irvine, this site brings together and organizes many Internet resources that can be used for finding positions, internships, teaching opportunities, and summers abroad.

2. **International and Overseas Job Search** – (www.overseasjobs.com)
    A constantly growing collection of over 600 unique links to employment resources in over 40 countries and regions intended to help you in your international job search.

3. **Jobstreet.com** – (www.jobstreet.com)
    Listing of job opportunities in Asia.

4. **Job Web Australia** – (www.seek.com.au)
    Australian on-line careers, resume, job search and recruitment service for companies and individuals.

There are several sites devoted to helping minority students with job opportunities, including:

1. **Black Collegian Online** – (www.black-collegian.com)
   An electronic version of the 28 year old, national career opportunities magazine. In addition to the abundance of career planning /job search information, there is commentary by leading African-American writers, lifestyle/entertainment features, general information on college life, and news of what is happening on college campuses today.

2. **Hispanic Jobs.com** – (Hispanic-jobs.com)
   Jobs for bilingual and Spanish-speaking professionals. Job seekers register free.

If you decide that you'd miss school too much if you graduated and immediately started working in a job, here are some sites that help you put off that decision for several years while you attend graduate school.

1. **NaceWeb** – (www.naceweb.com)

2. **A Guide for Candidates** – (monstertrak.monster.com)

3. **The Princeton Review** – (www.review.com/)

CHAPTER FOUR - PREPARING YOUR RESUME

"READ is a 4 letter Word!"

I'm sure you've heard the line, "You only have one chance to make a first impression." In a job interview, you have a chance to make two. Many students assume their first impression is made when the employer makes eye contact with them for the first time in the opening of an interview. However, in most cases, your first impression is made well before the interview. Your resume usually makes your first impression for you. To most employers, your resume is the most important application material (other application materials include a cover letter, references, application form, etc.). It holds the majority of the responsibility of getting you an interview. This should illustrate the importance of creating a resume that will stand out in a positive way. It's equally important to understand that while the resume is very important in getting you a job *interview*, it plays a smaller role in actually getting you a job. Once you're in the interview, it's not about your resume anymore. It's about you. Your resume got you there, but it is your personality and your communication skills that get you the job.

When creating your resume, you can't rely solely on your qualifications to get you an interview. The resume format is crucial, too. Employers spend less time reviewing and evaluate a resume less favorably if it utilizes a poor format or is too bulky. On average, employers spend less than four minutes reviewing your resume, so time is of the essence and you want them to be able to see the MOST important information about you in that limited amount of time. That means no matter how "pretty" you are as an applicant, if your resume is not designed well, you're going to be out of luck (and an interview). Remember that the format is going to be noticed by the employer before the content. A poor format in a resume would be like showing up for a job interview in sweats and a t-shirt. That would be one strike against you before you even open your mouth!

There is more than one way to create a good resume. The following tips come straight from the mouths of employers that interview college students for a living, so following their advice is one good way.

<u>**General Tips for Writing Resumes**</u>

1. Make it easy to scan! Employers don't want to have to read (yeah, I realize scan is a 4 letter word, too) your resume. If you had 100 resumes on your desk, you'd be looking for shortcuts, too. Scanning is key. The bullet format (which will be discussed shortly) is an effective way to create a resume that's easy to scan.

2. Make it concise! Rarely is there the need for a student to use more than one page for his or her resume. If you go onto a second page, usually, changing font sizes or margins shrinks it to one page. Back to that stack of 100 resumes in #1, any resume that flowed onto a second page makes your employer work even harder. Bad idea!

3. Make it relevant! Include the areas on your resume (i.e. education, work experience, activities, computer skills, etc.) you're most qualified in and that are relevant to the position for which you're applying for. Lead with your strengths!

4. Make it straightforward! Don't try to glamorize what can't be glamorized. All work experience is good even if it's not directly related to the job you're applying for, so don't worry so much about menial tasks in work experience. Don't try to make job responsibilities sound more significant than they really are. If you held a job as a server at a restaurant or a clerical worker for a company, you can list that on your resume but don't try to "milk" four responsibilities on four separate lines of your resume out of those jobs. While it's nice that you answered the phones as a clerical worker, that doesn't deserve its own line. One responsibility that includes three or four things is appropriate in this situation.

Example:

ABC Company
Clerical Worker
- Handled typing and filing duties, and answered and directed phone calls

As you can see in the example, it only took one line to mention several "less" significant responsibilities.

5. **PROOFREAD!** When you're finished, proofread it again. Have other people read it and check for errors, too. It's often difficult for you to catch your own errors. You don't want the employer to be the first person to notice an error on your resume. That wouldn't exactly help your rating on attention to detail.

6. Make it look professional! That means the printing, not the actual writing. You can use a professional looking word software, ideally one that allows you a little flexibility with format, font size, etc. Typewriters are unacceptable (if you can even find one).

7. Use 8.5" x 11" (#16-25) paper. Trying to use anything fancier can backfire and it's not worth the risk. The best colors to use are white or off white (beige, ivory, cream, etc.) Off white tends to distinguish the boldface print, which is the reason you use it in the first place. Print your resume, cover letter, and references are all on the same color paper. Also, use a large enough envelope so you don't have to fold your resume. It's also a good idea to type the address on the envelope (labels are OK).

8. Make your resume employer-centered, not self-centered! Tailor your resume to the specific job you're applying for.

9. Use boldface print and indentions strategically. Realize the more you use them, the less they stand out.

10. Vary the font size for emphasis. You can use size 16 or 14 for your name. The headings can be size 12 and the rest of the information can be 12 or 10 depending on how it fills up the page. Remember, you want it to fill up the page with margins no larger than one inch at the top and bottom.

11. Don't be afraid to use the vocabulary associated with the industry or position as long as you aren't doing it just to "show off." In fact, if you know the company you're applying to scans resumes into a computer and searches based on key words, it's a good idea to use these key words when creating your resume. The key words are often the relevant skills the position you're applying for requires.

12. Be ready to explain obvious gaps in work history. Employers don't expect you to work during the school year, but they do expect you to be gainfully employed during the summer unless you are attending summer school.

13. NEVER include names, titles, addresses or phone numbers of references on your actual resume.

14. Leave margins of 1/2" to 1" on the top and bottom margins and 1" on the left and right margins of your resume. Anything larger than an inch can make a employer think you don't have *enough* qualifications.

15. Do NOT include an introduction or concluding statement on your resume indicating why the recruiter should hire you. Most employers don't like them from students.

16. Create the information yourself. There are a lot of books out there that teach you how to write resumes, but a lot of them actually do you a disservice (even though you might not agree). They provide numerous examples of resumes for a variety of jobs. Many students have openly admitted to me that they pulled information straight from the book and put it on their resume. Write your resume yourself. Your resume is based on your experience and there should be no bigger expert on that subject material than you. Plus, when you get to the interview, it's much easier to talk about information on your resume if it's something you created about your own past experiences and not someone else's. It's extremely important to be yourself on a resume and in an interview.

There are two main types of resumes: Chronological and Functional. A third type, Skills, will be discussed but be wary of its down side. You'll be provided with several examples of each type, which are intended to serve as models to follow. Make sure you express your original ideas within the frameworks given. This will prove to be very beneficial when you sit down with an employer in an interview. Each section of the resume will be covered throughout the remaining pages of this chapter. Include the sections in your resume if they are relevant to you. It is important to avoid any unlawful data on your resume (i.e. age, sex, race, religion, national origin, marital status, health, etc.) Only include personal information relevant to the position (bona fide occupational qualifications). Any other information you choose to include could be used against you in an unlawful manner.

You will need to make several versions of your resume if you're applying for different types of jobs. As a general rule, you should only include personal information beginning with college. There are certain exceptions to this rule (i.e. Eagle Scout), but not many. Being an honor student in high school was nice at the time, but it doesn't set you apart anymore.

PARTS OF A RESUME

A. Name

Most employers suggest your name be in bold print, centered and in a larger font size (a size 14 usually works well) than the rest of the information on your resume. This allows your name to stand out. If there is one thing that you want the employer to remember after reviewing your resume, it's your name. Hopefully, with the rest of the information on the page, your name will conjure up a positive association in the mind of the employer.

B. Address(es)

If you only have one address where you want to be reached, center it under your name. If you have a campus address and a more "permanent" address (i.e. parents), put your campus address flush with the left margin, and place your permanent address (flush on the left side) over to the right margin. Make sure you include your area code and phone number for each address. You should also include your e-mail address. Make your address a smaller font size than your name and don't put it in bold.

EXAMPLES:

MICHAEL P. RUNNING
2217 Homel Lane
Springfield, CA 92154
(714) 365-2783
Email: Running@atl.com

BRUCE D. ADAMS

Present Address:
1424 Alexandria Dr.
Boston, MA 36544
(635) 625-2653
Email: BA@keol.com

Permanent Address:
1927 Chatsworth Blvd.
Portland, OR 36547
(867) 925-4623

Notice in both examples that the name is in **bold** but the address isn't. You need to be strategic when choosing which information to bold. Remember that, on average, employers spend less than four minutes reviewing your resume, so bold the information that you want them to notice first. Your name would be included on that list, but your address and phone number would not. The line below the address is mainly for aesthetic purposes as it serves to separate the name and address from the body of the resume. Employers like it but you don't have to use it.

C. Objective

Employers vary in their opinions on whether or not you should include an objective on your resume. The choice is yours. There are two key situations when an objective is particularly helpful. First, if you're giving copies of your resume to employers without a copy of a cover letter, include an objective. A job fair is an example of this scenario. At a job fair your goal is to mass market yourself, passing out as many resumes and meeting (i.e. schmoozing) as many recruiters as possible. When an employer sits down after a job fair and looks at your resume, she may not remember what it was that you wanted to do for the company. In this case, an objective would be helpful.

The second situation that calls for including an objective is when you're having difficulty finding enough information to fill the page. You don't want to have two-inch margins at the top and bottom of your resume because an employer will take one look at it and think you are under-qualified. It's imperative that the first impression of your resume is positive.

There are several reasons for leaving an objective off your resume. First, if you're sending a cover letter along with your resume, the same information in an objective is repeated in the first paragraph. You don't want to regurgitate information from your resume in your cover letter. Second, if you try and write a "career" objective, it's difficult to realistically sum up in one sentence what you want to achieve in a career that could last 30 to 40 years?

Third, your objective is not verifiable information. Employers, if they really wanted to and had the time, could check to determine if the information on your resume (with the exception of your objective) is truthful. They can contact past employers, request a diploma, transcripts, and other materials to ensure the information is accurate. It's very difficult to verify an objective. Even if the employer feels an objective is unrealistic for a particular applicant (i.e. a person with no college degree or work experience states in an objective he wants an upper level management position), it still represents the position that person wants. Ironically, the objective is really the only *subjective* piece of information on a resume. Finally, the objective is not really ordinal (rank order) data. All other information on your resume can be ranked against other applicants. For example, your education can be ranked on the basis of the school you attended, your GPA, your degree, etc. Your work experience can be ranked in terms of the quantity and quality of past jobs. Your honors, activities, computer skills and achievements can also be ranked based on their significance and relevance. While an objective could be ranked, it would be on less significant factors (i.e. clarity, grammar, well-defined, etc.) than those mentioned above. The final decision of whether or not to include an objective is up to you.

If you choose to include an objective, you need to determine how specific you want it to be. If you're too specific you might take yourself out of the running for other desirable positions with a company you're interested in. If you're too general you might be communicating to the organization that you're not sure what you want to do. The best thing to do is determine the necessary degree of specificity based on what you want to do. If you just want to get your foot in the door with a particular company and would be willing to do just about anything to work there, be general. If there's one and only one position you're interested in at a particular company, be specific and tailor your objective to meet the qualifications of that position. The best way to do that is include the skills, abilities, knowledge, etc. listed in the job advertisement.

EXAMPLES:

General: An entry level position in sales
Specific: A position in financial aid administration requiring excellent management and organizational skills and in-depth knowledge of financial aid procedures

In the "specific" example, the job advertisement mentioned the company was looking for a person with strong management and organizational skills. In the body of the resume, your goal would be to emphasize the areas you focused on in the objective (i.e. management and organizational skills) by providing specific examples. This is particularly easy with the specific objective because you stated exactly which skills you wanted to highlight. You don't need to include a period at the end of your objective because it's not a complete sentence since it doesn't contain a pronoun.

D. Education

Many employers feel that you should list education first on your resume even if you have a significant amount of relevant work experience (i.e. internships). Others feel you could lead with your work experience if it was more impressive than your education (i.e. several internships doing relevant work with reputable companies). The choice is ultimately up to you, but I would recommend leading with your education.

When listing education, only include colleges you attended and received a degree from. If you went to a four-year school and transferred after one year, don't list that school on your resume. This is designed to protect you. Employers could make the projection that based on the fact you switched schools in the past you'll be more likely to switch companies in the future. You can list a community college if you received an Associate's degree. Make sure you have your college stand out by putting it in bold print and all capital letters. Don't underestimate how prestigious your university might be to employers. Include the city and state after the university, but don't put them in bold.

Spell out your degree and list your major(s) and minor (if you have one), and date of graduation. Don't say "expected" or "anticipated" graduation. This makes an employer think you're not sure when you'll finish. If your plans do change, you can let the company know. Don't list the years you attended college because it's potentially unlawful information to the employer and can be used to discriminate based on age. You may have started college ten years ago and left and are now coming back to finish. Your dates would show you are older than the typical student and that could be used against you. You can list an "emphasis," an area you had several classes in that are relevant to the job you're seeking.

Your GPA is another area of debate. You're not required to list it on your resume, but you'd better be prepared to explain to an employer why it's missing (if you even get an interview). Accept responsibility for a low GPA and don't blame it on external factors ("my professor was mean," "his exams were unfair"). Some students have lower GPAs because they go to school AND have a job outside of school at the same time. Other students realized (maybe a little too late) that the major they chose when they started college wasn't right for them. These are both potential rationale for a lower GPA that most employers will understand (which doesn't mean they won't count it as a strike against you, but they might be sympathetic). A good rule of thumb is that if your GPA is over a 3.00/4.00, list it. If your GPA is below a 3.00, you're kind of in a no-win situation. If you list it, you may not get an interview because it's low. If you don't list it, you may not get an interview because the employer *thinks* it is low.

You should also list your GPA in the classes in your major if it's over a 3.00/4.00. Don't be fooled into thinking a high major GPA will excuse a low overall GPA. Many employers feel the overall GPA is more important than the major GPA. This might seem surprising since you'd think the employers would be more concerned with how you fared in a class relevant to your future job than one that isn't. However, if you really think about it, it's not all that surprising because employers are so intent on hiring well-rounded applicants (personally and academically). If you do well in the classes in your major but not outside of your major, employers might draw the conclusion that you'll do well in the job duties you're interested in but not do well in the aspects of your job you're less interested in. If you're listing both GPAs, make sure you indicate which one is overall and which one is in your major. Some employers have said they throw away a student's resume if no GPA is listed. That might not be fair (or legal), but it's something you should keep in mind. Don't forget to include an index after your GPA (3.19/**4.00**), because not all schools operate on the same scale.

EXAMPLE:

The following example probably applies to very few students since it lists two majors, a minor and an emphasis. However, the overall format is a good one to follow. Include which parts apply to you.

FLORIDA STATE UNIVERSITY, Tallassee, FL
    Bachelor of Science Degree
    Majors: English and Political Science
    Minor: Spanish
    Emphasis in Public Relations
    Graduation Date: May 2007
    GPA: 3.22/4.00 (Overall), 3.08/4.00 (English)
          3.19/4.00 (Political Science), 3.25/4.00 (Spanish)

Notice how the rest of the information under the name of the university is indented. Indentions, like bold face print, are another good way to make information stand out. Since people read left to right (last time I checked), the information furthest to the left will be read first. Using indentions and bold face print will highlight your most significant qualifications.

E. Significant Coursework

Employers like the idea of having you include relevant coursework on your resume. Your relevant coursework includes the classes you took in college that provided you with skills, abilities and knowledge relevant to the job you're applying for. Don't abbreviate the course. List the entire course title.

Now is a good time to mention the bullet format. Employers strongly prefer the bullet format to any other format. This isn't surprising considering the importance they place on being able to scan resumes for the most important information. The bullet format allows them to quickly scan the list of courses. You may use bold dashes instead of bullets because they achieve the same goal and they aren't so noticeable. Some bullets are so big and bold you notice them more than the information they precede. Compare the two following examples. The "good" one is preferred by employers and is much easier to scan for key information. The "bad" example is in a paragraph format and is much more difficult to scan. With the paragraph format, the employer must carefully read the information to see each specific course listed or one might be missed.

GOOD EXAMPLE: For a speech communication major's resume interested in a sales position

SIGNIFICANT
COURSEWORK
    - Professional Selling   - Interpersonal Communication
    - Theories of Persuasion   - Nonverbal Communication
    - Public Speaking   - Consumer Behavior

BAD EXAMPLE:

SIGNIFICANT COURSEWORK
Professional Selling, Theories of Persuasion, Public Speaking, Interpersonal Communication, Nonverbal Communication, Consumer Behavior

Notice how the courses mentioned in each example are the same, but in the bad example it would be easy for the recruiter to overlook a course. Nonverbal Communication is listed in the middle of a paragraph, which makes it easier to miss. The Consumer Behavior course begins at the end of one line and ends at the beginning of another line, which makes an employer work harder to scan the class.

F. The Body of the Resume

This is where the two main types of resumes mentioned earlier (chronological and functional) differ. Each focuses on a different aspect and will be covered separately. You need to determine which style best suits your qualifications.

CHRONOLOGICAL

If there is such a thing as a traditional resume style, this is it. In the chronological format (Figures 4.2, 4.3 and 4.4), you structure your information based on your work experience. List your jobs (starting with college) in reverse chronological order even if a position you held several jobs ago would be more impressive. If you worked while you attended college to help put yourself through school, include that on your resume (put it on the same line as the heading "Work Experience").

Next, you need to determine which of the following would be more impressive to the employer, the company you worked for or the position you held, and lead with that information. If you were a clerical worker at a top consulting firm, lead with the company (unless you desire a career in clerical work). If you were a manager at a company most people outside of its employees have never heard of, lead with the position. Put whichever you choose to lead with in all capital letters and bold print for emphasis. If you worked in a job that didn't really help you with relevant skills, but sure did make it a fun and enjoyable summer, you're in luck. Most employers feel that ANY work experience is good and should be listed on your resume, although the emphasis should still be on relevancy. Any job you held throughout college shows responsibility on your part whether or not your job duties were relevant to your intended career. It also shows teamwork, communication skills, etc. Since it's important to fit all of your qualifications on one page, less relevant work experience can be deleted if you're having trouble doing this.

You also need to include the dates for each job you worked. The key question is where to put them? Most employers prefer the dates of employment listed on the far left side of your resume. Make sure you list the month and year you started and finished a job ("present" is an acceptable ending date for a job you're currently working in). It sounds more professional if you list the names of the months (May-August) instead of the season (Summer). Employers like to see that you worked continuously, at least during the summers, while you were going to college. Any gaps you have in the dates of your work experience will most likely need explanation. You don't have to work during the school year, because employers know you're going to school full time. They do expect you to work during the summer unless you're taking summer classes. It's not the best idea to take some time off and relax after you graduate or during the summer, because employers like to hire students with a strong work ethic and that's not a good way of projecting one.

For each job you list, include your major responsibilities or duties. The specific number depends on the job. If you worked in a job where you had four significant duties, list them. If you worked in a job where you only had one main duty, you can limit your list to one. Don't try and glamorize your duties on paper as most employers can see right through that strategy. The focus, as it should be throughout the entire resume, is on responsibilities relevant to the job you are seeking. However, if you worked in a job where all of your responsibilities were menial, you can group them together. For example, let's say you answered phones, filed charts and entered data into the computer in your job as a clerical worker. Listing each of those on a separate line makes it look like you're trying to make them seem more significant than they really are. Instead, group them together and list as one responsibility. Begin each responsibility with an action verb, past tense if it is a past job and present tense if it's a job you're still working in. By doing that, even if the employer only scanned the first word of each responsibility, she would still be able to see an example of a skill. The employer could scan the entire responsibility to discover the context in which the skill was performed, but she wouldn't have to. Capitalize the first letter. To help you with a list of action verbs, see Figure 4.1 in the appendix. Here are some good examples of work experience:

TOP CONSULTING FIRM, City, State
Clerical Worker

May – December, 2006
- Filed purchase orders and answered phones
- Assisted in the processing of customer orders
- Conducted equipment inventory weekly

MARKETING MANAGER
Joe's Bar and Grill, Bellview, TX

5/05 – Present
- Design and direct advertising campaign
- Coordinate marketing strategy for all computer software
- Trained sales staff of 22 people on new marketing strategy

Notice in both examples that indentions are used so the company (in the first job) and the position (in the second job) stand out more because they are farther to the left and in bold print and all capital letters. In the marketing manager position, there are two responsibilities that take up a second line. In that case, don't use another dash. Align the first word of the second line right below the first capitalized letter on the first line (not directly beneath the dash). Both jobs also utilize the bullet format for listing responsibilities. That is strongly recommended as it makes each individual responsibility stand out more. Here is an example of a work experience listing without indentions and the bullet format:

        TOP CONSULTING FIRM, City, State
        Clerical Worker
        Filed purchase orders and answered phones, assisted in the processing of customer orders, conducted equipment inventory weekly

I'm sure you can determine that if an employer is scanning your work experience, this example make it more difficult to distinguish each individual responsibility than in the good examples listed above. Employers think work experience is the most important area on your resume in determining whether or not you get a job interview. This should emphasize the importance of interviewing for internships during college.

FUNCTIONAL

This type of resume (Figures 4.5 and 4.6) focuses more on your abilities and achievements than your work experience. It pulls examples from many different areas of your college experience (i.e. classes, work experience, extracurricular activities, etc.) and groups them under the headings "Capabilities" and "Achievements."

CAPABILITIES
- Organized weekly meetings and study sessions for fraternity
- Delivered speeches to high school students on the importance of staying in school
- Supervised student athletes during nightly study tables
- Met every paper and project deadline during college
- Assessed customer needs and matched with products

ACHIEVEMENTS
- Selected "Outstanding Student in Senior Class"
- Voted "Most Dedicated Student" by the University faculty
- Wrote an article on diversity training that was published in an international management journal
- Achieved an overall grade point average of 3.96/4.00

The capabilities listed provide examples of organizational, communication and leadership skills along with the ability to meet deadlines. The achievements all represent accomplishments that would be significant to employers. If you have space, you can still list work experience after this section, but most people using this type of resume list capabilities and achievements instead of detailed work experience. You can see how work experience was still incorporated into the resume as two job responsibilities (supervising study tables and assign customer needs) were listed below capabilities.

SKILLS

This type of resume (Figure 4.7) groups your information into relevant skill sets. It's usually used by students that don't have a lot of work experience. A word of warning before choosing this type: Some employers evaluate a skills' resume lower because, to them, it indicates the student has little relevant work experience (which you know by now is a very important determinant in your job search success). If you do choose this type, use the heading "Relevant Skills" and then determine several broad categories for your skills. The categories should represent skills that would be desirable for an employee to possess in the job you're seeking. The simplest way to accomplish this is to look at the job advertisement as it usually lists relevant skills the job seeker should possess. Communication, Organizational, Leadership and Analytical are examples of relevant skills. After each heading, list several (the specific number is dependent on your experience) examples of times in your past you have demonstrated the relevant skill. Similar to the functional resume, you can pull these examples from classes you have taken, activities you were involved in during college, or from "less" relevant jobs in which you have worked.

RELEVANT SKILLS
Communication
- Lectured to elementary school children on the dangers of drug use
- Delivered informational and persuasive speeches to 30 students
- Wrote a daily column for the school newspaper, *The Informant*

You could still list your work experience without the detail if you don't feel it's relevant. List the organization, city, state and date of employment. You don't have to put any of the information pertaining to the job in bold print if you don't think it is relevant enough to draw the employer's attention to it right away.

WORK EXPERIENCE
1/06 – Present   Midway Auto Center, Lafayette, IN
9/05 – 12/05   Newstat Marketing, Coley, IL

G. Activities and Honors

List your activities and honors from college. Any involvement in campus organizations or membership in honor societies is good to include. You can, but you do not have to list the dates at the end of each honor or activity. It's best to only include membership in a social organization (i.e. fraternity or sorority) if you emphasize a leadership position you held. You may be thinking, "I've always listed my fraternity or sorority before without indicating the positions I held and no one has ever said anything to me?" Membership in social organizations is not an area that employers normally probe into, because membership isn't an essential requirement for performing a job. This is designed to protect you, not hurt you. There are a lot of employers that don't hold positive opinions of social organizations, so it's not worth taking a chance by listing one. If you did hold a leadership position; however, the focus changes to the fact that leadership skills are relevant to the position you're seeking.

You have two choices when organizing your activities and honors. First, you can list them under separate headings. This is a good idea if you have three or more of each.

ACTIVITIES

Vice President, Business Student Association
Standards Committee, Beta Theta Sigma Fraternity
Treasurer, Kappa Beta Phi Business Fraternity
Head Resident Advisor, Rydel University

HONORS

Recipient, Rydel University Academic Scholarship
Dean's List (3.5/4.0), 5 semesters
Awarded "Top Business Student Award"
Chosen from 75 students to represent Rodell in National Symposium

You can also group them under one heading if you only have one or two of each.

ACTIVITIES/HONORS

Vice President, Business Student Association
Treasurer, Kappa Beta Phi Business Fraternity
Recipient, Rydel University Academic Scholarship
Awarded "Top Business Student Award"

H. Volunteer Work

Any volunteer job or community service you perform during your college years is great to include on your resume because they show prospective employers you are interested in other things besides money. Even if you know in your heart that you want the big bucks more than anything else in a job, keep this hidden from the employer. If a volunteer job was extensive enough to be listed under work experience, you can do that. If not, put all of your volunteer and community service under one heading.

VOLUNTEER WORK
    Habitat for Disadvantaged People
    Binfurd County Animal Shelter
    Broadview Shore Convalescent Hospital

You don't want or need to give a detailed description of your honors, activities or volunteer work (you can do this in a cover letter). By trying to include too much information (i.e. descriptions) in your volunteer work brings up the problem of making the employer read, not scan, your resume. These represent good topic areas of conversation, so be prepared to discuss them. It's actually a good strategy on your part to only list the organization so the employer will want to ask you about what you did. It's a lot easier question to answer than a lot of other options that are at the employer's disposal.

I. Computer Skills

With the boom of the information age, students with strong computer skills are becoming increasingly more valuable to employers. The question is, what computer skills make you "strong?" If you know how to surf the Internet, send e-mails and use Microsoft Word, you probably are not what employers would consider a person with strong computer skills. Most employers believe you should ALWAYS list computer skills, even if the job you're applying for doesn't require them to start. Just because a job doesn't require computer skills now doesn't mean it'll always be that way. When computer skills do become a necessity, and chances are very good they will, the company wants you to already have a good working knowledge. They don't want to have to spend the time and money involved in training you. When you list computer skills on your resume, avoid a phrase like "familiar with…" because it doesn't mean anything specific. To be familiar with a piece of software may indicate you know it so well you could have written the program yourself or that you are aware that there is software by that name but you don't know the first thing about using it. You might very well be tested on your computer skills if the job for which you are applying requires extensive computer knowledge and ability.

COMPUTER SKILLS
    Hardware: IBM, Macintosh
    Software: MS Word, Windows XP, Windows NT, and PowerPoint
    Languages: HTML, Java, C++

J. Interests/Hobbies

Employers are split on whether or not to list your interests and hobbies on your resume. This leaves the choice up to you. I'd only list interests if you have listed everything else about yourself you possibly can and your margins at the top and bottom of the page are still too big. There is the possibility that your interests may give the employer some common ground on which to start the interview ("I love to jump out of airplanes, too"). Unfortunately, the potential downside to listing them is that employers often feel the only reason you're including them on your resume is because you don't have enough other relevant qualifications.

INTERESTS
    Hot air ballooning, fencing, spelunking, roller hockey

K. Foreign Languages

If the job you're applying for requires the ability to speak or write a foreign language, include that on your resume. Listing foreign languages you can speak or write that aren't required could potentially be used to discriminate against you based on national origin. You'd be much better off listing information that is more relevant to the position. Your two best choices for words to use when listing your languages are "Fluent" and "Conversational." Fluent implies that you converse, read, and write the particular language extremely well. If that's not the case, you're better off writing that you're conversational in the language. That gives employers the impression that you could hold your own in the conversation, but not that you're ready to take up residence in the country in which the language is primary.

FOREIGN LANGUAGES
    Conversational in French and Spanish

The areas mentioned so far aren't your only choices of information you can include on your resume; however, they do tend to be the most common for students in particular. If you don't have strengths in one of the areas mentioned, don't list it. If you have strengths in an area not mentioned, you can include that instead. The key is to fill up your one page resume without making it look too bulky.

L. References

It used to be a given that the line "references available upon request" found its way on to the bottom of a resume as the very last line. Most employers feel that you don't need to include it anymore because it wastes valuables space and it's assumed. This line is the one piece of information on your resume that can't speak for itself. An employer can make an impression about you based on every other line of your resume without talking to anyone. Unfortunately, though, for that last line to actually say something about you, the employer would have to get your list of references and call the people on the list. Considering how many applicants they're usually dealing with, that's a very time consuming proposition. Most employers don't contact your references very often anyway, because they figure you're smart enough to only include a reference who is going to give you rave reviews. When you stop to think about it, how does it really help the employer to hear positive things about every applicant? It doesn't help any *one* applicant really stand out. Now if a reference has negative things to say, that's another story. You probably can kiss your chances goodbye!

Just because you didn't include that last line doesn't mean you can leave your list of references at home. Make sure you bring a list of references to every interview. When you're choosing your references (i.e. the people who don't have anything on you and are guaranteed to say great things), keep several things in mind:

1. Employers feel they make the best references, meaning your former or current employers are their preferred choice for you to list. Professors would be the second best option. It's a good idea to have at least one of each. Other choices include friends, family members, priests, co-workers, etc. If you're wondering why friends and family members possess so little popularity as references, the answer is simple: Employers know your mommy and daddy are only going to say good things about you. As a general rule, if you're thinking about using a reference that has the same last name as you, think again. You can use a relative, but only if you're using that person as a professional reference (i.e. he or she was your boss). Unfortunately, the propensity for ANY reference to say nice things about you is a problem. Employers don't contact your references that often because they don't really get the quality information that helps separate applicants in the hiring process.

2. Make sure you ask your references for permission even if you are 100% sure They'll agree. It's a matter of consideration and it also allows them to be prepared in case they're contacted. You should provided them with any information you have (i.e. resumes, cover letters, etc.) that would help them say nice things about you to potential future employers. A resume will help your reference comment on the "well rounded" you and not just how you did in one job, one class, etc.

3. Type your references on a separate sheet of paper from your resume. Use the same kind and color paper as your resume and cover letter.

4. The specific number of references you choose to include is up to you. You should have a minimum of three. You can have up to five or six if you want. Try to use people from different areas of your life. Including one or two current or former employers and one or two professors is a good idea.

5. Include each reference's name, position (if a professional reference), street address, e-mail address and phone number. The phone number is probably most important to include because most employers will call, not write your references. They don't want pen pals. They want to get the information about you ASAP. You can list your references in alphabetical order or in order of significance. A current or former employer in the same industry as the company you're applying to would be an ideal reference to list first.

6. Don't forget to include your name on your list of references, because it could get separated from your application materials and it would be difficult to match it back with your resume and cover letter without your name at the top.

References for Betty L. Wilkins

Dr. Ruben Amaro
Professor
Seal State University
Department of Communication
San Diego, CA 92536
(619) 746-2635
Ruben_Amaro@Seal.Edu

Mrs. Martha Knox
President
Shuler and Associates
1145 Chatsworth Blvd.
San Diego, CA 92534
(619) 857-3624
MK@sand.com

Dr. Rebecca L. Ferguson
Assistant Professor
Department of Psychology
Seal State University
San Diego, CA 92536
(619) 746-8746
Rebecca_Ferguson@Seal.Edu

ELECTRONIC RESUMES

Electronic resumes have become an extremely popular method of sending copies of your resume to employer. The main reasons are that they are received much faster, they are less expensive and you can send your resume to hundreds of employers all over the world at the touch of a button. An electronic resume is one that can be sent to a company via the Internet or e-mail. This method allows companies to electronically scan resumes once they are received and store this information in a database. Employers can electronically screen resumes in the database based on specific criteria that they have determined. This criteria is often relevant skills, abilities, personality characteristics, etc. Due to the increasing number of electronic resumes received, it's a good idea to also send a hard copy of your resume the good old-fashioned way for those jobs you really want. I've talked to a lot of employers that are actually steering away from electronic resumes because they're being inundated with them. It's so easy for students to send them that tons of unqualified applicants are sending them because they figure "What do I have to lose?" That doesn't mean you shouldn't send one, just realize a hard copy would be a wise idea, too.

When you're creating an electronic resume, it's a good idea to utilize a format that you KNOW every organization will be able to read. Unfortunately, Word and WordPerfect do not fall into that category. The safest format to use is a text format, with the most common text format being ASCII (American Standard Code for Information Interchange). There are certain things to keep in mind when creating an electronic resume, including:

1. Bullets are not recommended because they are not recognized by the text format. Use dashes or asterisks instead. In fact, graphics of any kind should be avoided.

2. Use all capital letters in your resume for the purpose of emphasizing certain information (i.e. headings, your university, etc.). Underlining and bold face print will not be recognized in a text format. If you had a line going all the way across the page below your name and address, you can use a series of dashes to accomplish the same purpose.

3. Emphasize key words in your resume so it will be included when a key word search is conducted to find the most qualified resumes.

4. When you are saving your resume, use the "Save As" command and save as an ASCII file. Make sure your extension to the file is .txt (resume.txt).

5. When you are finished with your resume and you have saved it as a text file, it is a good idea to send it to someone you know (that does not work for the company you want to work for) so he or she can see how it looks after the file is opened.

6. If you are sending a cover letter, too, make sure and include both your resume and cover letter in the same file. You can accomplish this by copying and and pasting your cover letter in front of your resume in the same file.

7. In the subject matter area of the e-mail, use the position title for which you are applying or you can use a reference number if one is supplied.

8. It is becoming more common for companies to provide an on-line form that you can use to paste your resume directly into the company's web site.

9. Many on-line job search sites offer you a resume building service that, in most cases, is free of charge. This service will take you step by step through the process of creating a resume that is appropriate for an on-line search.

10. The same rules for following up with a company after you send a resume and cover letter in the mail apply when you send them on-line. It's a good idea to follow up with a phone call about a week after you send them. Your purpose for contacting them (as far as the company is concerned) is to check to see if they received your application materials. You and I both know, however, that you're really calling in hopes of getting a chance to speak with someone in a hiring position so you can make a good impression.

If you want more information on sending electronic resumes, check out the following web site:

EResumes and Resources - (http://www.eresumes.com)

Or you can read the book:

Electronic Resumes and Online Networking, by Career Press

As you can see in this chapter, there are numerous ways to format and send a resume. There is not just one right way to write a resume, but there are some ways that are definitely better than others. The example resumes in this chapter have been employer tested and approved. The 188 employers that responded to the survey evaluated them and their suggestions were incorporated into the final versions you see. This should indicate to you that if you follow the guidelines discussed in this chapter your resume will be evaluated positively, at least in terms of its format, and that creates a good first impression about you in the eyes of the employer. That is a key first step on the path to success in the job search.

Figure 4.1 – **Action Verbs** (make present tense if currently working in the job)

Accomplished	Designed	Maintained	Revised
Achieved	Determined	Managed	Scheduled
Acquired	Developed	Marketed	Selected
Adapted	Directed	Measured	Served
Administered	Documented	Mediated	Set up
Analyzed	Earned	Minimized	Settled
Anticipated	Edited	Modified	Showed
Applied	Educated	Monitored	Sold
Approved	Employed	Motivated	Solved
Arranged	Enforced	Negotiated	Sponsored
Assisted	Established	Observed	Staffed
Assigned	Estimated	Obtained	Standardized
Attained	Evaluated	Operated	Started
Audited	Examined	Organized	Stimulated
Broadened	Expanded	Originated	Strengthened
Built	Financed	Oversaw	Structured
Calculated	Formed	Participated	Studied
Clarified	Formulated	Planned	Suggested
Collaborated	Founded	Prepared	Supervised
Communicated	Fulfilled	Presented	Supported
Completed	Generated	Prevented	Tailored
Conceived	Guided	Produced	Taught
Concluded	Handled	Programmed	Terminated
Conducted	Helped	Promoted	Tested
Consolidated	Headed	Proposed	Trained
Constructed	Hired	Published	Transferred
Consulted	Identified	Recommended	Translated
Contributed	Implemented	Recruited	Trimmed
Controlled	Improved	Reduced	Tripled
Converted	Increased	Re-established	Uncovered
Coordinated	Influenced	Reinforced	Undertook
Corrected	Initiated	Reorganized	Unified
Created	Instituted	Reported	Used
Decreased	Instructed	Represented	Utilized
Defined	Integrated	Researched	Verified
Delivered	Interviewed	Resolved	Won
Demonstrated	Invented	Revamped	Worked
Designated	Investigated	Reviewed	Wrote

Figure 4.2 - **Chronological Resume without an objective**

DEREK D. ARNOLD
1234 Toledo Avenue
Lubbock, TX 72047
(635) 938-2645
Email: DDA@ptb.com

EDUCATION

TEXAS TECH UNIVERSITY, Lubbock, TX
Bachelor of Business Administration Degree
Major: Marketing
Emphasis in Public Relations
Graduation: May 2007
GPA: 3.23/4.00 (Overall)

RELEVANT COURSEWORK

- Theories of Public Relations   - Public Speaking
- Marketing Principles   - Sales Techniques

WORK EXPERIENCE   (Financed 75% of college expenses)

PUBLIC RELATIONS INTERN
KYEW RADIO - 1090 AM, Lubbock, TX

May 2005- Present
- Coordinate studio production and live radio broadcast
- Assist sales personnel in selling airtime to local businesses
- Promote radio station at local community and sporting events

ANCHOR/REPORTER
KNSD-TV Channel 42, Lubbock TX

August 2004 - May 2005
- Anchored a 30-minute newscast five days a week
- Gained experience in writing and directing newscasts
- Reported for a five minute news segment around the city
- Edited and wrote news packages
- Produced an educational game show "Double Trouble"

COUNSELOR
Kanomo Camp, Brussells, MO

May 2004 - August 2004
- Counseled children between the ages of 10 and 15
- Incorporated Christian principles with discipline of athletics
- Taught dance, swimming, wind surfing and sailing

COMPUTER SKILLS

Hardware: IBM, Macintosh
Software: Windows XP, Excel, Word, WordPerfect, PowerPoint
Languages: HTML, Java, and C++

ACTIVITIES/HONORS

Public Relations Committee, Student Foundation
Chaplain, Kappa Gamma Theta Sorority
Recipient, NCT Communication Scholarship
Dean's List (3.5/4.0), 3 semesters

Figure 4.3 - **Chronological resume with an objective**

WILLIAM K. MONTGOMERY

Present Address
1526 Chatsworth Blvd.
Ann Arbor, MI 40284
(354) 746-2640
Email: Monty@umich.edu

Permanent Address
2075 Willow Drive
San Diego, CA 92106
(619) 223-0147

OBJECTIVE

A position in sales requiring strong communication, organizational and persuasive skills

EDUCATION

UNIVERSITY OF MICHIGAN, Ann Arbor, MI
Bachelor of Arts Degree
Majors: Speech Communication and Psychology
Graduation: August 2007
GPA: 3.19/4.00 (Overall), 3.21/4.00 (Psychology)

WORK EXPERIENCE

SALES REPRESENTATIVE
Lakeway Rent-a-Car, Ann Arbor, MI

May 2006 - August 2006
- Opened and developed the market segment and sources of business offering the best potential
- Developed, serviced and maintained customer base
- Prepared all sales correspondence and promotional literature
- Represented the company in local community events
- Planned and coordinated internal employee programs for generating leads

SALES REPRESENTATIVE
Southern Federal Bank, Ann Arbor, MI

August 2005 – May 2006
- Implemented successful marketing plan to assure maximum exposure to targeted real estate offices
- Explained programs to customers to ensure a quick, efficient loan process
- Created a marketing strategy including making sales calls, mailings and telemarketing

COMPUTER SKILLS

Hardware: IBM and Macintosh
Software: Windows NT, Microsoft Access, Novell, and Oracle

VOLUNTEER WORK

The Humane Club of the United States
Barney's Buddies, a youth camp for underprivileged children

FOREIGN LANGUAGES

Conversational in Spanish

Figure 4.4 - **Chronological resume with a Master's degree**

TRACY ANNE FILLER
7564 Poinsettia Way
Seattle, WA 53423
(746) 365-2735
Email: KJP@dem.com

EDUCATION

Graduate   **UNIVERSITY OF WASHINGTON**, Seattle WA
                        Master of Business Administration Degree
                        Major: Finance
                        Graduation: December 2006
                        GPA: 3.47/4.00

Undergraduate   **UNIVERSITY OF OREGON**, Eugene, OR
                        Master of Arts Degree
                        Major: Political Science
                        Graduation: May 2004
                        GPA: 3.24/4.00 (Overall), 3.19/4.00 (Major)

WORK EXPERIENCE

                        ACCOUNTING CLERK
                        Buckfield Home Corporation, Seattle, WA
May 2005 - August 2005   - Promoted to Land Development Department in one month
                        - Handled lot purchases and processing of relevant paperwork

                        LOAN PROCESSOR
                        ILD Mortgage 4Corporation, Seattle, WA
May 2004 - August 2004   - Processed FHA, VA and Convention loans from origination
                           through funding
                        - Reviewed mortgage risk, real estate appraisals and evaluated
                           the borrower's credit worthiness

                        PEPE'S WORLD FAMOUS MEXICAN RESTAURANT, Eugene, OR
                        Server
January 2004 - May 2004   - Helped customers with meal selections, cleaned tables and
                           answered phones

ACTIVITIES

                        Vice President, University of Washington Graduate Student Association
                        Treasurer, University of Washington Graduate Student Association
                        Head Resident Advisor, University of Washington

COMPUTER SKILLS

                        Hardware: IBM, Macintosh
                        Software: MS Excel, MS Access, MS Word and MS PowerPoint

VOLUNTEER WORK

                        Adopt-A-School program (University High School)
                        Volunteer, Woodway Family Center

Figure 4.5 - **Functional Resume with an Associate's Degree and an objective**

KASSIE KEMP

Present Address	Permanent Address
645 Fifth Street	53462 Jewel Avenue
Columbia, SC 35242	Manhattan, KS 26354
(824) 847-4635	(534) 824-2830
Email: Salsa@leo.com	

OBJECTIVE

A managerial position requiring strong leadership, organizational and communication skills

EDUCATION

UNIVERSITY OF SOUTH CAROLINA, Columbia, SC
Bachelor of Science Degree
Major: Sociology
Minor: Management
Graduation: May 2006
GPA: 3.08/4.00 (Overall), 3.31/4.00 (Major)

COLUMBIA COMMUNITY COLLEGE, Columbia, SC
Associate Degree in Political Science
Graduation: May 2004
GPA: 3.35/4.00

CAPABILITIES

Interviewed and made recommendations of prospective employees
Created and designed displays for produce department at Chuck's
Delivered speeches to high school students on achieving their potential
Resolved customer problems or complaints quickly and respectfully

ACHIEVEMENTS

Department accounts for one million dollars in annual sales
Initiated department cost reduction plan lowering overhead 25%
Devised individual stocking system for $8,000 per week sales category
Achieved an overall GPA of 3.31/4.00 while working 40 hours per week

WORK EXPERIENCE

5/05-8/05	Chuck's Grocery Store, Columbia, SC
12/04 – 5/05	Hathaway's Department Store, Columbia, SC

ACTIVITIES

Youth basketball coach, Springdale Elementary School
Wedgeway High School tutor in math

FOREIGN LANGUAGES

Fluent in Japanese

INTERESTS

Horseback riding, golfing, croquet and country dancing

Figure 4.6 - **Functional resume with work experience**

JORDAN KEMP
5342 East Blvd.
Chicago, IL 46359
(847) 947-3652
Jkemp@bigdog.com

EDUCATION

        UNIVERSITY OF CHICAGO, Chicago, IL
            Bachelor of Science Degree
            Major: Electrical Engineering
            Minor: Mathematics
            Graduation: May 2006
            GPA: 3.02/4.00 (Overall), 3.45/4.00 (Major)

CAPABILITIES

        Designed and built a microprocessor for departmental use
        Formulated a plan for manufacturing system design
        Developed a computer system to assist in micrographic projections
        Presented engineering plans to employees and management

ACHIEVEMENTS

        Selected Top Engineering Student at the University of Chicago for the
            2005 academic year
        Awarded University of Chicago Alumni Scholarship
        Member of Tri Sigma Engineering Honor Society
        Elected President of Tri Sigma Engineering Honor Society

WORK EXPERIENCE

        ENGINEER ASSISTANT
            Dynamic Electronics, Chicago, IL
August 2005 - Present   - Assist Head Engineer with projects and plans for model
              construction
            - Design micro processing chip for computer system
            - Develop a circuitry board and training program for interns

VOLUNTEER EXPERIENCE

        Whispering Pines Convalescent Hospital
        Windy City Animal Shelter
        Great Lakes Marathon for Education

ACTIVITIES

        Beta Theta Kappa Engineering Society
        Member of Society of Professional Engineers (SPE)

COMPUTER SKILLS

        Hardware: IBM, Macintosh
        Software: MS Word, Windows XP, WordPerfect, and PowerBuilder
        Languages: C++, Visual C++, HTML, C, Java

Figure 4.7 - **Skills Resume with an objective**

SCOTT A. WICK

Present Address	Permanent Address
6453 Southside Drive	756 West 34th Street
Salt Lake City, UT 36544	Minneapolis, MN 76453
(534) 645-7364	(987) 276-3562
Email: Downy@utah.edu	

OBJECTIVE

A position in law enforcement requiring strong leadership and communication skills

EDUCATION

UNIVERSITY OF UTAH, Salt Lake City, UT
Bachelor of Arts Degree
Major: Criminal Psychology
Minor: Sociology
Graduation: December 2006

SIGNIFICANT COURSEWORK

- Social Problems   - White Collar Crime
- Criminology   - Sociology of Law
- Criminal Justice   - Child Abuse and Neglect
- Social Deviance and Control   - Racial and Ethnic Diversity

RELEVANT SKILLS

Leadership
- Organized pre-practice offensive drills for teammates
- Advised players on handling relationships with coaches
- Led group members in preparing for class presentations
- Coordinated charity drive for the Darby Days fund raiser

Communication
- Lectured on drug abuse to elementary school children
- Delivered speeches to 30 students in a public speaking course
- Participated in numerous group oral presentations during college
- Wrote weekly sports column for the school newspaper

ACTIVITIES/HONORS

Athletic scholarship for five years at the University of Utah
Three-year starter at offensive tackle
Scout team player of the week four times during freshman year
Member of the Walk for Brotherhood fund raising drive

COMPUTER SKILLS

Hardware: IBM, Macintosh
Software: MS Word, Windows XP, Windows NT, and PowerPoint

INTERESTS

Golfing, running, basketball, chess, wrestling

CHAPTER FIVE - PREPARING YOUR COVER LETTER

"Got you Covered!"

Chapter four mentioned that employers don't want to spend a lot of time looking at your resume. Well, they want to spend even less time looking at your cover letter. You probably felt cheated when you learned spend less than four minutes evaluating your resume, especially considering the amount of time it took you to write it. Compared to the cover letter, that'll seem like an eternity. On average, employers spend about two minutes reading your cover letter. If you combine the time employers spend scanning your resume and reading your cover letter, you get a whopping six minutes. That's probably pretty discouraging considering all of the effort you put into each.

Just because employers only spend two minutes reading your cover letter, doesn't mean you don't have to write one. It's still a good idea to send a cover letter along with most of the resumes you send out. There are times when you don't need a cover letter (i.e. job fair, posting a resume on a web site, etc.), but if you're unsure whether you need one or not, send one. You're better off sending one and not having the employer read it than not sending one and leaving the employer wondering where it is. The cover letter does provide you with the opportunity to explain what you are interested in doing for the company and to sell yourself to the job. The more it looks like your qualifications match those of the ideal applicant for the position, the more likely you will be to get an interview.

General Tips for Writing Cover Letters

1. Don't make your cover letter more than one page. If you're concise and to the point, this shouldn't pose a problem. Longer isn't better when it comes to cover letters. Remember, you're not the only person sending a cover letter for the employer to read. You may have taken essay tests in college where you figured the more you wrote, the more likely you were to actually say something of value. That strategy doesn't work with employers when it comes to cover letters.

2. Sell yourself to the job description. The job description will normally include a list of the necessary knowledge, skills and abilities (KSA's) the person applying for the position must possess. Provide examples and illustrations (unique from anything on your resume) from your past when you have demonstrated those KSA's. Use action verbs like you did in your resume.

3. Proofread very carefully for typographical, grammatical and spelling errors. One little error could go a long way in showing the employer that your attention to detail is lacking. Have someone else proofread it too, since it's often difficult for you to find your own errors.

4. Don't just regurgitate information that's on your resume. This is your chance to expand on what you feel (and hopefully what the employer will feel) is most

important. You can expand on areas listed on your resume, but don't simply repeat them. For example, if you listed relevant coursework on your resume, discuss in your cover letter, projects or presentations that were a part of the class that would demonstrate relevant KSA's.

5. Try and address each cover letter to the specific person who will be reading it. If you don't know the specific person's name, give the company a call and try to find out. This phone call also gives you the chance to get your name into the minds of members of the organization. Make sure you are professional, polite and considerate no matter whom you speak with. Even if the person you speak with has little direct decision-making power in you landing a job, he or she may have a lot of influence with the people who do. Don't use "To Whom it May Concern" at the beginning of a cover letter because chances are good it won't end up concerning anyone.

6. Focus on what you can do for the company, not what the company can do for you. You can mention the benefits you'll gain from them, especially if you're interested in an internship, however, the focus on the letter should be what you can do for them. It's a good idea to discuss facts about the company you learned through your research that are impressive (a smart way of patting the company on the back) to you.

7. Send an original and creative cover letter with each resume. Don't try to create a universal cover letter.

8. Be enthusiastic about yourself, the position and organization.

9. Be confident. You can demonstrate confidence in your writing style. Don't write statements like "I feel I would be an excellent addition to XYZ company." It sounds like you are trying to defend yourself. Instead write, "I would be an excellent addition to XYZ company."

10. Type on bond paper (the same kind and color as your resume).

11. If you mention in the closing paragraph that you'll contact the employer, don't forget to do that. I doubt you will. If you wait five to seven days, that should be plenty of time for the employer to receive and review your application materials. It doesn't hurt to take the initiative and be aggressive.

The following pages will include example paragraphs you can use in the opening of a cover letter (Figure 5.1) along with a framework for cover letters (Figure 5.2) in general, and three sample cover letters to use (Figures 5.3 – 5.5) as guides when creating your own.

Figure 5.1 – **Example Opening Paragraphs for Cover Letters**

1. **Response to an Advertisement** – You are sending a cover letter to a company in response to an advertisement listing an available position. With an advertised position, your chances improve considerably.

EXAMPLE: I am writing to you about the Park Ranger position advertised in the Sunday, March 21st edition of the Hickory Gazette. I am very interested in the position and my knowledge, skills and experience would be invaluable to XYZ company and the management of your forest.

2. **Using a Contact** – You are sending a cover letter to a company hoping to be considered for an interview (for an advertised or unadvertised position). You are using the name of a connection that the company would recognize and make a positive association with.

EXAMPLE: Betty Robell, a close family friend, referred your name to me. I am graduating in December with a Master's Degree in Accounting. I am interested in pursuing a career in corporate accounting and would appreciate the opportunity to interview with ABC Accounting Firm for a full time position. (If the position was advertised, refer to the specific position and source).

3. **Prospecting** – You are sending a cover letter to a company hoping it has an opening, but nothing is advertised. This is a very difficult way to get an interview.

EXAMPLE: I am a senior in Human Resource Management graduating in May. I am interested in XYZ Company because of its emphasis on teamwork and commitment to excellence. My knowledge, experience and drive to succeed would make me an excellent addition to your company. I am looking to start in an entry-level position.

Figure 5.2 – **Standard Format for a Cover Letter**

Your Street Address
City, State, Zip Code
Date

Name of the person to whom you are writing
Title of the person
Name of the organization for which he/she works
Address of the company

Dear Dr./Mr./Mrs./Ms./Miss _____:

First Paragraph: State the specific purpose for writing. There are several possible reasons for writing (Figure 5.1) and the appropriate one should be brought to the recruiter's attention immediately. Also, mention the specific position for which you are applying, and how you found out about the opening (friend, placement center, newspaper, etc.). This is the place to mention a connection. A connection can make a huge difference in whether or not you get an interview or a job offer.

Second Paragraph: Discuss how your background (i.e. school, work, activities, etc.) qualifies you for the position you are applying for without restating what you wrote on your resume. This is your chance to go into more depth and provide the recruiter with examples and details of your experiences. You may want to discuss what knowledge and skills your classes, projects or extracurricular activities taught you. You can also list accomplishments you have achieved at school or work. Then, move into a discussion of how your past work experience qualifies you for the position. Mention how certain tasks from past jobs would be directly applicable to tasks in the job you are seeking. Point out again that you worked WHILE attending college and what this taught you (i.e. time management, responsibility, etc.). If you have extensive work experience, divide the second paragraph into two separate, smaller paragraphs. Focus one paragraph on your academic experience and focus the other paragraph on your work experience.

Final Paragraph: Indicate your desire for an interview and let the individual know you are willing to meet on his or her terms (as far as the time and place). Close by restating your phone number and mention you will contact the recruiter. Do not rely on the recruiter to contact you. Show some initiative. The recruiter is far too busy to call back every applicant. You might be sitting by the phone for a long time if you wait for the recruiter to call.

Sincerely,

(Signature)

Name (Typed)

Enclosure: Resume and References (If you send references)

Figure 5.3 – **Sample Cover Letter (Advertised Position)**

5684 West 32nd St. #243
Chapel Hill, NC 87512
June 25th, 2006

Mr. James Riley
Personnel Manager
Top Automobile Manufacturer
3040 Round Road
Detroit, Michigan 48095

Dear Mr. Riley:

I am writing in response to your recent advertisement for a customer sales representative that was listed in the Sunday, June 24th edition of the Detroit Press. With my strong people skills and determined work ethic, I would make an excellent addition to Top Automobile Manufacturer.

I admire your company's vision to be the worldwide leader in the automotive industry and to be committed to providing "total customer enthusiasm" through your employees. I have the ability to generate customer enthusiasm. I recently completed an extensive 40 hour mediation course. This training provided me with the skills necessary to analytically separate fact from emotion in order to manage disputes and resolve conflicts in the most effective, practical, and positive manner possible. These techniques are essential in satisfying the interests of the customer while still protecting the interests of Top Automobile Manufacturer.

In accordance with the mediation certificate, my academic studies at The University of North Carolina in Communication and Organizational Management have given me a diverse curriculum that relates directly to your organizational culture. I am referring to your dedicated approach to strengthening employee commitment by actively facilitating, teamwork, leadership, and motivation. I am certain that, by incorporating my own knowledge and experience from these same areas, I can guide employees toward achieving common objectives. I can also assist in the improvement of customer relations in order to meet and exceed customer expectations.

The enclosed resume is a summary of my qualifications, training and experience. I will be in the Detroit Metroplex area the second week of July. I would be honored to interview with you while in Detroit. I will call you the first of next week to try to schedule an interview date. Please feel free to contact me at (815) 468-2534 if you have any questions or need any additional information. I look forward to meeting with you.

Sincerely,
(Signature)
Jillana E. George
Enclosure: Resume

Figure 5.4 – Sample Cover Letter (Unadvertised Position – Prospecting)

1800 Primrose
La Jolla, CA 92156
April 15, 2006

Mr. Roger Hanson
Director of Sales Staff
MJB Electronics
5326 Patton Drive
Memphis, TN 64534

Dear Mr. Hanson:

I am a senior Business Administration major graduating in August. My varied work experience in business, coupled with my education, has prepared me for a career in management. I am interested in working for MJB Electronics because of its commitment to its employees and the community. I would like to start in a management trainee position.

While at San Diego State University, I had numerous opportunities to demonstrate my management and leadership skills. I served as the inaugural leadership chairperson for my sorority. As leadership chairperson, I implemented a leadership and mentoring program that pinpointed members' strengths and directed them in paths that best fit their interests. I have also been given several leadership and management tasks within my sorority including a special project head in a theatrical program. In this role I organized a task team, delegated responsibilities, and promoted enthusiasm among the members.

Along with my college experience, I have had the opportunity to work as a salesperson with several companies. This exposure has been instrumental in strengthening my interpersonal communication skills and flexibility in approaching and dealing with diversity in people and the workplace.

I would welcome the chance to speak with you about my qualifications and starting a career with MJB Electronics. I will contact you the week of April 22nd to follow up on this letter. If it is more convenient for you, I can be reached at (619) 222-4653. Thank you for your time and consideration.

Sincerely,

(Signature)

Wendy E. Hall

Enclosure: Resume and References

Figure 5.5 – Sample Cover Letter (Advertised Position with a Contact)

7253 Hatfield Avenue Apt. #3
Natchfield, IL 54635
October 14, 2006

Mr. Peter Filler
Director of Personnel
Top Marketing Firm
Washington Hall
934 North Washington
Seattle, WA 64537

Dear Mr. Filler:

I am writing you regarding my interest in the marketing representative position advertised in the Sunday, October 7th edition of *The Examiner*. Eric Fife, the northwest district manager suggested I write to you personally. My knowledge and experience in the marketing field would make me an excellent addition to Top Marketing Firm.

I have had the opportunity to work in a marketing department in two different jobs at both the collegiate and professional level. Interacting with players, coaches and the media on a daily basis has served to strengthen my communication skills and teach me different perspectives involved in the sports' industry. Through my experience with the Cleveland Blizzard and the West Central State University football program, I have been able to gain insight into the efficient operations of two different organizations.

Along with my experience, I have found the time and dedication to actively belong to numerous campus organizations, holding offices in several of them. Besides working and being involved on campus, I have maintained a GPA of 3.7/4.0, which shows organizational and time management skills. My excellent oral and written communication skills, dedication, work ethic, and enthusiasm would make me an asset to the Fort Sound Rocketeers' organization.

I would welcome the opportunity to speak with you about the marketing assistant position. I will contact you the week of October 21st to follow up. If it more convenient for you, I can be reached at (928) 263-3746. Thank you for your time and consideration.

Sincerely,

(Signature)

Spencer R. Koltun

Enclosure: Resume

CHAPTER SIX - RESEARCHING PRIOR TO YOUR INTERVIEW

"Read all about it!"

Imagine you landed your first big job interview with THE company you've wanted to work for all along. This is the job at the very top of your list. In the days leading up to the interview, you spend countless hours preparing for every possible question you might be asked. The big day arrives. You're dressed for success. You arrive at the interview site 15 minutes early (a good recommendation to follow). The employer greets you at the door. You make great eye contact as you smile and shake her hand firmly. You respond enthusiastically when asked how your day is going. You're feeling good and ready for anything that might come your way. You're brimming with confidence as the employer poses her first question, "Tell me what you know about our company." Your jaw drops. Nausea, not confidence, is now brimming in your stomach. You spent so much time preparing for questions about yourself you failed to prepare for questions about the company. You think to yourself "I know how important being honest in the interview is, this is a good time to start." With that in mind you respond, "I haven't really had time to research it yet." Any positive impression you made on the employer is long gone the moment you finish your reply. You've just told the employer, without actually verbalizing it, that you really don't care about her company.

I'm sure you've spent countless hours researching for papers and studying for exams during your college career. That preparation is very important in determining how well you perform in your classes. However, the most important preparation you'll do isn't designed to help you with a paper or an exam. It will help you start your career (i.e. your life after college). Employers feel researching the company and position is the most important step in preparing for your interview. Common sense might tell you that the best thing you can do is to know how to act or how to answer questions about yourself. While that is very important, knowing about the company and job is even more important. The reason is that your knowledge of the company and position is, in the employer's mind, a direct correlation of how interested you are in the company and position. The more you know, the better off you'll be. Telling the employer, "I think your headquarters is located in St. Louis, but I'm not sure" isn't going to impress her. So what should you find out about the company and position? While it would be impossible to know everything, there are areas to find information about in your research.

Areas to research about the company

- Products and services (know what the company does)
- Geographic locations (plant, headquarters, offices, etc.)
- Company history (past performance, reputation)
- Company philosophy (goals, values, mission, etc.)
- Future plans (trends, expansion, merger and acquisition plans, etc.)
- Major competitors (know who they are)
- Financial status (stock performance, sales volume, short/long term profit picture)
- Size (number of employees)

- Management (know the CEO, President, etc.)
- Know everything on file at your campus placement center

Areas to research about the position

- Responsibilities and duties (know what you would do in the position)
- Advancement opportunities (career paths in the field)
- Relocation policies
- Training program
- Skills required
- Education (required degrees, certifications, etc.)
- Salary and benefits (let the employer initiate this topic)
- Type of supervision
- Travel (quantity involved)
- Turnover rate

Where do you find this information so you can provide it when asked? Usually, you don't have to go any further than your computer. The internet is an excellent source of information and most companies have a web site. Typing www.companyname.com (or .org) or using a search engine with the company name will usually allow you to find the site you're looking for.

You can also visit your Career Service Center where you should find a searchable on-line database that you can use to search companies by name. Many on campus Career Service Centers have libraries where they keep printed literature and brochures from companies that interview on campus, too.

Try your library (you should know where it is by now) and talk to the reference librarian about how to use services like Lexis-Nexis or InfoTrac. Libraries also usually contain online resources or services from major business publications like *The Wall Street Journal*, *Kiplinger's*, *Forbes*, *Fortune*, *Money*, or *Investor's Business Daily*. You can use bizjournals.com to find business news by industry and/or location. Haven't read enough book yet in college? Here are some more that deal with information on companies.

- Corporate Yellow Book
- Dun and Bradstreet's Million Dollar Directory
- Dun and Bradstreet's Directory of Service Companies
- Dun and Bradstreet's Career Guide
- Dun and Bradstreet's Reference Book of American Business
- Dun and Bradstreet's Who Owns Whom
- Encyclopedia of Associations
- Encyclopedia of Business Information Sources
- Everybody's Business: An Almanac
- Hoover's Handbook of Private Companies

- Hoover's Handbook of American Business
- National Job Bank, 2005
- Notable Corporate Chronologies
- Standard and Poor's Corporation Records
- Standard and Poor's Register of Corporations, Directors, and Executives
- Standard and Poor's 500 Directory
- Thomas' Register of American Manufacturers
- U.S. Industry Profiles: The Leading 100
- Ward's Business Directory of U.S. Private and Public Companies

If you want information on companies outside of the United States, try:

- America's Corporate Families and International Affiliates
- Directory of American Firms Operating in Foreign Countries
- Dun and Bradstreet's Europa
- Encyclopedia of Associations: International Organizations
- Hoover's Handbook of World Business
- International Directories of Company Histories
- Marconi's International Register
- Moody's International Manual
- The… Global 500 Directory
- Ward's Business Directory: Major International Companies
- Who Owns Whom – United Kingdom and the Republic of Ireland; Continental Europe
- Yearbook of International Organizations

If you're looking for financial or statistical information on companies, check:

- Industry Norms and Key Business Ratios
- Market Guide
- Market Profile Analysis
- Market Share Reporter
- Moody's Handbook of Common Stocks
- Moody's Manuals
- Morningstar Mutual Fund Sourcebook
- Standard and Poor's Corporation Records
- Standard and Poor's/Lipper Mutual Fund Profile
- Value Line Investment Survey

Another key step in your preparation is to know yourself better than you ever have before. This is best accomplished by conducting the self-analysis described in chapter three. If you don't think about questions you might be asked about yourself, you'll soon realize you don't know yourself as well as you thought you did.

Another recommendation for preparing for interviews is to practice, practice, practice! The notion "practice makes perfect" applies to the job interview. Although perfection may be an unattainable goal, you should strive for as close to perfection as possible. Adequate practice will enable you to perform much better during the interview. Part of your practice should include thinking about answers to possible questions you might be asked. Don't try to memorize answers. If you adopt this strategy, it's as if you are reciting lines from a script during your interview and most employers will see right through this act and questions your honesty and sincerity.

It's a very good idea to call or stop by your school's placement center to see if it offers any type of mock interviews. Mock interviews are set up for students to gain valuable interviewing experience in a "practice" setting. Real recruiters from real companies usually conduct the interviews. The biggest difference is that there isn't a real job on the line (although it is important to realize that if you really do impress the recruiter with your personality, skills and experience, it can lead to a real interview at a later date.) You should prepare for this interview just like you would for an interview where a real job is at stake. If you don't take it as seriously by not preparing as much as you normally would, you won't really benefit from the experience as much as you should.

At the end of every interview, whether it's real or practice, assess your performance. Obviously the recruiter will be doing this too, but you can learn a lot from analyzing how you did. What did you do well? In what areas do you need improvement? What questions gave you trouble? It's normal to struggle with some questions when asked of you the first time because you can't prepare for every one. If you struggle with the same question a second time, you have no one else to blame but yourself. Even if your school doesn't offer mock interviews, practice in front of family and friends and get their feedback. Of course, be careful answering practicing in front of mom, because mom's normally think everything their kid's say is great. While this may be an ego booster, it may not be the kind of objective criticism you need.

It's helpful to talk with other students who have been through interviews to get their advice. Ask them what employers were looking for, what types of questions were asked, etc. If you view every interviewing experience as an opportunity to learn, you'll find yourself more comfortable and able to fare better in future interviews. Chapters seven and eight provide you with an extensive list of possible job interview questions. Look through as many of them as you can and think about what you would say if you were asked each question. Focus on areas you would discuss, not on a verbatim response. If you have an outline and not a script in your head, it will be much easier to appear natural when responding to employers.

Once you have researched all of the necessary areas prior to the interview, it's time to begin preparing for the employer's questions.

CHAPTER SEVEN - BEHAVIORAL INTERVIEWING

"Be on Your Best Behavior!"

This may be the most important chapter in the book because it covers an interview strategy that many companies have turned to in order to help them make a more effective decision in predicting your ability to do the job. It's also going to make your life in the interview more difficult. It's called Behavioral Interviewing and it's based on the premise that the best predictor of future performance is past performance. For lack of a more creative title, I'll refer to the questions students have been asked for years and years in interviews as "traditional" questions. An example traditional interview question that has stood the test of time is "What is your greatest weakness?" Let's see, I hate people, I like to kick puppies, I like to set things on fire... Yeah, right, like you'd really say any of those things even if they were true.

Instead, you'd try and be a little craftier and say that you're a perfectionist, or that you try and take on too much responsibility at one time. If you try one of these answers, an employer with any experience will see right through you. In her mind (and yours, too), a perfectionist is someone that takes more time to complete his or her tasks to ensure things are done perfectly and won't have to be redone. Is that a crime? Heck no, that's a good thing and you know it, but employers know you know it.

Two additional problems with traditional questions are that students tell employers what they think they want to hear (i.e. the socially acceptable answer) and students can tell a flat out lie, and the employer may never know. Let's say an employer asks you to tell her what characteristics make a good leader. You could say "a good leader is someone who is motivated, shows initiative, and has excellent listening skills." Not bad, that would be some good stuff. The only problem is that the employer didn't ask you what leadership skills *you* have. You may have so little motivation that you have a hard time dragging yourself out from under the covers for a class that starts at noon. You may be a person that waits for someone else in your group to tell you what to do, thus showing no initiative whatsoever. You might also be a chronic daydreamer, so when other people are talking, you're in your happy place. As you can see, there are clear deficiencies with traditional interview questions.

That's where behavioral interviewing comes into the picture. A behavioral question forces a student to tell a story about something factual from his or her past that illustrates a skill, ability or personality characteristic the employer is looking for. These questions make it a lot more difficult to just make something up or tell the employer what you think she wants to hear. For some sample behavioral questions, see the list at the end of this chapter.

In a survey of 197 organizations, 84% (165) indicated they currently ask at least one behavioral question in an interview. Of those, 46% (91) said that EVERY question they ask in an interview is a behavioral question. I'd say these findings are pretty convincing. How do you prepare for a behavioral interview? Here are two good steps to follow:

1. Try and determine the relevant behaviors that a recruiter is going to ask you about. Look at the job advertisement (if there is one), because it normally mentions relevant skills, abilities and personality characteristics a company is searching for.

EXAMPLE:

VanDelay Industries is currently seeking a part-time employee (less than 30 hours per week). The qualified candidate should be a self-motivated, organized, individual with excellent interpersonal and communication skills, and the ability to work well as part of a team.

This ad mentions several skills and abilities: motivation, organization, interpersonal, communication and teamwork. These would be a good place to start. I'll cover how to prepare an answer shortly, but what do you do if there is no job advertisement? Use your common sense and that big gray mass between your ears. There are a lot of skills, abilities and personality characteristics that are relevant to just about EVERY job available, regardless of whether or not they're mentioned specifically. The job advertisement I listed did not mention leadership skills. Does that mean you don't need to be a leader and be prepared to answer a question about your leadership skills? No way, be ready! Employers want to hire leaders, period! Even if you're starting at the bottom rung on the organizational ladder, they still want you to possess strong leadership skills. Why? Employers want you to move up in the organization, into jobs where leadership skills are essential. They don't want to find out then that you stink as a leader. Here is a list of skills, abilities and personality characteristics that are pretty common, regardless of the job. That means be ready to provide answers about all of them.

BEHAVIORS

- Adaptability
- Analytical
- Assertiveness
- Attention to detail
- Coaching
- Communication (oral and written)
- Control
- Conflict Management
- Continuous learner
- Coordinating
- Creativity
- Critical thinking
- Decision-making
- Delegation

- Listening
- Motivation
- Negotiation
- Organization
- Participative
- Perseverance
- Personal responsibility
- Persuasiveness
- Planning
- Practical learning
- Planning
- Presentation skills
- Professionalism
- Rapport building

- Energy
- Entrepeneurial
- Flexibility
- Goal setting
- Independence
- Influence
- Initiative
- Innovation
- Integrity
- Interpersonal
- Judgment
- Leadership
- Results oriented
- Resilience
- Risk taking
- Sales ability
- Sensitivity
- Teamwork
- Technical knowledge
- Technical proficiency
- Tenacity
- Time management
- Training
- Troubleshooting

2. Prepare an answer to a question. Now that you know the kinds of behaviors employers are looking for, how do you answer a question about one of them? The best thing to do is create outline in your head that covers the essential areas. Employers look for several things in an ideal answer to a behavioral question. A recommended structure would be to follow the STAR response technique. It covers Situation, Action, and Results. If you are wondering what the T is for, I'll explain it with the Action.

SITUATION

Describe the situation that you were in or the task that you needed to Accomplish. You need to describe a specific event or situation, not a general description (very common mistake) of what you have done in the past to demonstrate the skill, ability or personality characteristic they are asking about.

ACTION

The "T" can be included before you describe the action. Describe what was going through your mind (i.e. thoughts) at this point that led you to believe you needed to take some action. Then, describe the action you took and make sure the focus is on what YOU did. If you use an example from a group project from a class you took, describe your efforts, to the entire team. Give them specific, detailed description of what you did.

RESULTS

Describe the outcome. How did it all turn out? What happened as a result of your actions? What did you accomplish? What did you learn? Obviously, provide a positive example. If all of your group members stormed out of the room as a result of your actions, keep that to yourself and come up with a different example.

Make sure you give the employer enough detail so she understands. If you don't she is going to ask follow up questions, which means you are making her work harder to get the information. Bad idea! You want to make it as easy on her as possible by doing the LARGE majority of the talking. You can provide an example from a previous job, from a volunteer experience, from a class you were in, or any other relevant event. I would have two or three examples ready for each behavior in question. You want at least one situation that is completely positive, and one that started poorly, but either ended positively or you made the best out of the situation and learned from it.

I'm not a big fan of providing actual answers to job interview questions in a book because they makes students (not you, of course) more likely to "borrow" them for their own job interviews. Then an employer ends up hearing professionally prepared responses to their questions. Even having said all that, I'm still going to provide you with an answer for a behavioral question, because I think you need to see one to get a feel for what it sounds like. I mentioned earlier you want to make sure and be specific, not general. I'll provide you with a bad sample response first and then a good one. You should be able to tell the difference very easily.

BEHAVIORAL QUESTION:

Give me a specific example of a time you demonstrated your leadership skills.

BAD ANSWER

"There have been numerous times I have demonstrated my leadership skills. I was the team leader on a project in my last job and that role required me to constantly use my authority to solve problems in the group. Group members would often have personality conflicts and try to get by with the least amount of effort and not carry their fair share of the workload. I would have to step in and get them to stay focused and work harder so the group would be successful in accomplishing its goal."

If you begin your answer with "There have been numerous times…," you're off to a bad start. The employer asked for a specific example. Specific is not general, and it also doesn't mean brief. You can be very specific AND be very detailed at the same time. Be as detailed as you can be as long as you don't start babbling off on some unrelated subject.

GOOD ANSWER – Using the STAR approach

SITUATION

"I was the team leader of a project in my last job at XYZ consulting. We were working on a training project for ABC law firm. The project involved designing a computer software program that would allow ABC's lawyers to easily track and record their billable hours. The company had contacted us because its current recording system inundated them with paperwork and was fraught with errors. XYZ selected me to lead a team of five employees to assess ABC's current situation and design a software program that would alleviate the problem. The other four employees selected to be on the team with me were Scott, Frank, Abby and Karen. We first met Monday, January 14th and right away I could see that there were going to be some personality conflicts among the group. Scott and Abby got into a heated debate about the assessment techniques we would use to gather information on ABC's current system. Abby told Scott she was more qualified to give input on the topic due to the fact that she had once worked for a law firm and was more aware of the problem. Scott replied that he had been with XYZ longer and had more experience with assessment techniques. The entire time this was going on, Frank was sitting and reading the sports page while Karen was polishing her nails."

ACTION (with THOUGHTS included)

"My frustration level with the group was increasing. I knew the best thing to do for the good of the group would be to deal with these problems right away so we could avoid any lingering problems in the future. I took the initiative to confront Scott and Abby. I told Abby that while her work experience for a law firm would be very valuable to us, she also needed to recognize the value in Scott's experience with the company. I told Scott that his experience with the company would be a huge asset during the course of the project but he also needed to understand the perspective Abby could add since she was the only member of the group that had worked for a law firm in the past. I mentioned to Frank that I was a sports fan too, but there was a time and a place to read the sports page and this was not it. Karen quickly realized from my comments to the other three group members that her actions were also inappropriate so she put her fingernail polish away. I reiterated to the group the importance of the project and that management had demonstrated its confidence in us and that we should do everything possible to not let them down."

RESULTS

"My actions had a positive effect on the group. Abby and Scott both apologized to one another for the comments they had made. Abby told Scott she would be more open minded to his suggestions in the future knowing that they would be based on many years of experience. Scott said that he would avoid calling Abby names and listen to her suggestions and try to combine her knowledge of law with his experience with assessment techniques. Frank promised that he would leave the sports page at home from now on and come to future meetings prepared to work hard. While Karen did not say anything specific to the group, it was apparent from her body language and attentive, concerned look she would save her personal hygiene for home. The group meeting ended up being very productive, as did the project as a whole. Thanks to the groups' diligent effort and expertise, the software and training program we created is now in high demand by law firms from around the country."

I'd like to think you would have been able to figure out which answer was good and which answer was bad, even without the labels. It should have been easy to see the difference in specificity and depth between the two answers. It also might be quite an eye opener when you realize this was an answer to ONE question. Imagine if you are asked ten to fifteen behavioral questions. Pack your bags, you're going to be in the interview for awhile.

The odds are against you being able to avoid at least one behavioral question in an interview. As I mentioned earlier, roughly four out of five organizations will ask behavioral questions so you better get your situation, actions and outcomes ready.

You may be a little worried now that you've learned about behavioral questions and the challenge they present. Yes, they're more challenging to answer, but you can handle them. Have a story ready about each of the relevant skills, abilities and personality characteristics and try and make them from the past year or two. With all of this talk about behavioral questions, be careful not to forget about the old stand by, traditional questions. They're going to be included in most interviews, too. I would prepare for the interview as if both behavioral and traditional questions will be asked. That way, you'll be able to handle anything thrown at you. If you go to an interview with the mindset that only traditional questions will be asked and they zing behavioral questions your way, say goodbye to the job offer because you won't be getting it!

SAMPLE BEHAVIORAL QUESTIONS

1. Give an example of how you applied knowledge from one of your courses to a job.

2. Tell me about a time when you had to use your written communication skills to get your point across.

3. Describe a time when you were faced with a stressful situation that demonstrated your coping skills.

4. Give me an example of a time when your point of view was challenged. What did you do?

5. Tell me about a time when you were forced to make an unpopular decision.

6. Tell me about the last time someone at work or school misunderstood what you were attempting to communicate.

7. Describe the most creative thing you have ever done.

8. Give me an example of a time when you had to work with someone who was difficult to get along with. Why was this person difficult? How did you handle that person?

9. Tell me about a time you had to mediate a conflict.

10. Give me an example of a time you delegated a project effectively.

11. Tell me about a time when you missed an obvious solution to a problem.

12. Describe a situation that required a number of things to be done at the same time. How did you handle it? What was the result?

13. Tell me about a time your active listening skills really paid off.

14. Describe a time when you had to defend one of your decisions.

15. Describe a situation where you were able to overcome a personality conflict in order to get results.

16. Tell me about the last time you failed to meet a goal. How did you handle it?

17. Give me a specific example of something you did that helped build enthusiasm in others.

18. Give me an example of a time where you pushed yourself to do more than the minimum.

19. Tell me about a situation that could not have happened successfully without you being there.

20. Tell me about a difficult situation when it was necessary to keep a positive attitude. What did you do?

21. Describe a situation where it was important to display tact.

22. Tell me about a time when you had to use your presentation skills to influence someone's opinion.

23. Give me a specific example of a time when you used good judgment and logic in solving a problem.

24. Give me an example of a time when you motivated others.

25. Give me an example of a time you had to make a split second decision.

26. Tell me about a time when you had too many things to do and you had to prioritize your tasks.

27. Describe a time you had to lead people that did not want to be led.

28. Tell me about a time you were proactive in dealing with a problem.

29. Give me a specific example of a time you were an effective team player.

30. Tell me about a time when you felt you really went beyond the call of duty and had to take a risk. What did you do?

CHAPTER EIGHT - TRADITIONAL INTERVIEW QUESTIONS

"Oldies, but Goodies!"

As mentioned in chapter seven, even with the importance of behavioral interviewing, you still need to be ready for traditional questions. It would be virtually impossible to prepare for every possible behavioral and traditional question you might be asked in an interview. However, your job is to try! *Any* question, whether it's behavioral or traditional, you've prepared for will be a heck of a lot easier to answer than one you didn't prepare for. That's why preparation is so important. It's a wise idea for you to realize going into an interview that you will be asked questions you didn't prepare for. This will allow you to handle the situation better when it happens. Be calm, think for a second or two and then do the best you can. Try not to be thrown off by the fact that you were caught off guard. That will make it even more difficult to give a good answer.

Here are some questions to anticipate and prepare for before your next interview.

TRADITIONAL INTERVIEW QUESTIONS

1. Tell me about yourself.

This question needs some explanation because it's so popular and yet it's one of the more difficult questions you may be asked. In fact, you might think it's too broad to ask. Well, that's exactly why so many recruiters like to ask it. It's often the first question in an interview and it puts you on the spot early. Employers want to see how you handle pressure and this question does a good job of measuring that. Pressure is an inherent part of every job and employers want to know if you can handle it BEFORE you start working for their company. Don't ask employers "Could you be more specific?" or "What would you like to know?" They would have been more specific if they wanted to. They intentionally want to be general because it helps them assess your priorities and your communication skills. Don't pause for more than a few seconds (two or three) before responding because it communicates indecisiveness and uncertainty on your part. You need to prepare for this question in particular because it is one of the more difficult questions to answer if you do not.

How do you answer this question? You have several directions to take your answer. This is a lot of information to work with. One interesting thing to remember about this question is that there is not really a "right" answer. Many employers are not as concerned with what you say, but how you say it. This answer represents a great opportunity for an employer to assess how well you think on your feet (even while you are sitting down), which is an important quality to possess. When preparing an answer for this question, remember that the employer has most likely already seen your resume, so don't just regurgitate information from it. You can talk *about* information on your resume. Some students like to talk about their family and growing up. That's not a bad idea (even though I don't think it's the best option) because it does show family values, which are important to most companies. When you really think about this question, it represents an excellent opportunity to sell yourself. A great way of doing that would be to talk about skills, abilities, personality characteristics, training, experience, etc. that you possess that sets you apart from the other applicants. Provide examples of times you demonstrated them, thus making it more of a behavioral question. Whether you like it or not "Tell me about yourself" is here to stay, so you better be ready!

2. Why did you select to interview with our company?

3. What do you know about our company?

4. From your research on our company, what suggestions would you make?

5. Why should I hire you?

6. What can you do for us that someone else can't?

7. What or who has influenced your career choice the most?

8. Why did you select your college?

9. What did you learn from college outside of the classroom?

10. Describe your most rewarding college experience.

11. How has your college experience prepared you for your career?

12. Why did you choose your degree?

13. What classes did you enjoy the most/least and why?

14. Who was your favorite professor and why?

15. Is your GPA is an accurate reflection of how hard you worked in college? Why or why not?

16. Why is your GPA so low (or missing from your resume)?
    (Take responsibility and do not make excuses)

17. If you could, how would you plan your college career differently?

18. Describe your ideal job.

19. Define _____.
    - Could be just about anything: leadership, cooperation, teamwork, etc.
        (Do not give a dictionary definition, put in your own words)

20. Describe your leadership style.

21. How has your work experience prepared you for this position?

22. What work experience has been the most valuable to you and why?

23. How would others (friends, professors, employers) describe you?

24. What three words would you use to describe yourself?

25. What part do you normally play on a team? (Leader of course!)

26. What qualities do you possess that make you a good team player?

27. Tell me about your most significant (school, work, etc.) accomplishment.

28. What is your greatest weakness?
    - What are you doing to turn it into a strength?

29. What is your greatest strength?

30. Where do you see yourself in five or ten years?

31. What do you know about the position you're interviewing for?

32. What do you think is a fair salary?

33. Tell me what salary range you expect.

34. How much are you worth?

35. How important is money to you?

36. How do you evaluate success?

37. What challenges are you looking for in a position?

38. What is the last book you read?
    (If Dr. Seuss comes to mind, you're in trouble)

39. How do you keep abreast of current events?

40. What frustrates you? ("Nothing" might seem like a good answer, but it's not)

41. How do you handle criticism?

42. What is the biggest obstacle you have had to overcome?

43. What makes you angry?

44. How do you study?

45. Describe traits of people that annoy you.
    (Be careful not to describe the employer)

46. If you could change any world event, what would it be and why?

47. What scares you about our company?

48. What interests you least about this job?

49. What will your biggest initial challenge be in this job?

50. On what occasions are you tempted to lie?

51. What do you want to hear first, the good news or bad news and why?

52. Sell me _____.
    - Very popular in interviews for Sales' positions. Be ready to sell something to the employer on the spot (the pen you're holding in your hand is a good place to start)

Another trend that's becoming increasingly popular with employers is to ask questions that seemingly have no relevance to the job whatsoever. While you may be thinking that these questions would be unlawful, they're actually lawful questions because they do probe into some inherent aspects of most jobs (i.e. the ability to think on your feet, the ability to handle stress, your creativity, etc.). The key to answering one of these questions is not to let it rattle you.

1. If you had to be part of a _____ what part would you be and why?
   - It could be a hamburger, salad, pizza, etc. Go for the integral part (meat on the hamburger, lettuce in the salad, etc.). The other ingredients (i.e. cheese, tomatoes, croutons, etc.) play a supporting role. Employers want to hire leaders, not followers.

2. In the news story about your life, what would the headline say?

3. Why are manhole covers round?

4. If you only had six months to live, what would you do?

5. If you could compare yourself to any animal, which would it be and why?

6. If you had a job interview that was in a city 500 miles away, would you fly or drive and why?

7. If you were stranded on a deserted island, and all of your essential needs (i.e. food, water, shelter, etc.) were taken care of, what three things would you want with you?

8. If you could have dinner with anyone from history, who would it be and why?

9. If Hollywood made a movie about your life, who would play you?

10. If you could be any character in fiction, whom would you be and why?

11. If you could be a superhero, what would you want your superpowers to be?

Right now you're probably thinking "What the heck?" An employer that asks you one of these questions is usually as interested in how you handle yourself when answering the question than what you say in your response. If the question noticeably throws you for a loop, the employer might think you'll be less likely to handle an unexpected situation or circumstance that could arise on the job. Try to relax and do the best you can.

CHAPTER NINE - APPLICANT QUESTIONS

"Payback!"

You're probably just starting to come to grips with the daunting task facing you in preparing for all of the possible questions you might be asked. Well, get ready for more questions, but this time they're questions you get to ask the employer. Yeah, that's right, you're going to ask questions, too. Ah, payback time. This is actually a very important step in the interview process. If you think the only reason you're in an interview is to let the employer determine if you're the right person for her company, the boat's sailing and you're missed it. The interview process is a two way street. Get it engrained in your brain now that you're in an interview to make sure the job and company are right for you, too. The best way to do this is to ask good questions. You need to make sure and research the company and job first so you can develop some quality questions that you were NOT able to find through your research. Don't play the game of asking a question simply because you think it will impress the employer. If you make a practice of this, you'll get a lot of information that's not important to you and will not help in your decision-making.

ALWAYS come to an interview prepared to ask questions. It's not uncommon in a first interview with a company to not get to ask any questions, because, honestly, in first round interviews employers may not care about what you want to know. Allowing you the opportunity to ask questions takes time. That time might take away from their opportunity to interview additional applicants. The employer often keeps total control of a first round interview (letting you ask a question gives you some control). Employers may also ask every applicant in a first round interview the exact same questions, which makes comparing applicants easier and results in more reliable ratings. Even though you may not get a chance to ask questions in a first round interview, you need to have questions prepared for EVERY job interview.

You can look at your opportunity to ask questions as a right, something you should be able to do if you want to. If you feel like the employer is getting ready to end an interview without letting you ask questions, you can request time to ask questions if you feel comfortable doing so. In a first round interview you may not feel comfortable, but in second round interviews, I would definitely request the opportunity to ask questions if you don't feel it's in the employer's plans to let you. This is the only way to get information (in addition to what you learned through your research) about the company and position to help you decide if the job and company is right for you. Don't ask questions that you should have been able to discover in your research and don't ask questions about salary in a first round interview. Those questions tell an employer you lack professional maturity and common sense.

A good suggestion is to ask a question that PROVES to the employer that you have done your research. You will find some answers in your research, but certainly not all of them. Focus your questions on the undiscovered areas. It is also a good idea to ask questions that probe into the employer's opinions. That is a way of communicating to the employer that you respect her opinion and value her input and insight. One thing you need to realize before you ask an employer a question in an interview is she, too, might give you an answer that you want to hear. She might be unwilling to pass on any negative information about the job or company. If you do succeed in getting an office visit, ask the employees that are currently working in the position you're interviewing for how they feel about their company and jobs. They may be more likely to pass on the honest information, even if it is negative. Employers like to hear students ask the following questions:

GOOD QUESTIONS

1. What do you value about this company?

2. Why have you stayed with the company?

3. What would you change about the company?

4. What was your path to your current position?

5. What do you like best/least about your job?

6. What are the strengths and weakness of the company?
    - How is your company working to improve its weaknesses?

7. What are the strengths and weakness of your competitors?

8. What makes your company unique?

9. Tell me about your training program (ask about a part that was not included in your research).

10. What qualities are you looking for in new hires?

12. What are the most challenging aspects of the position?

13. What is the most important issue for the new hire to tackle?

14. What would you like done differently by the new hire in this position?

15. How does your company train and develop a new employee?

16. In six months, what would the successful candidate have accomplished?

17. What kind of mentoring and training style do you have?

18. Describe the opportunities for advancement in your company.

19. How does your company support people returning to school seeking advanced degrees?

20. What are the opportunities for personal growth?

21. How are employees evaluated and promoted?

22. How does the company recognize outstanding employees?

23. How are risk taking and creativity rewarded?

24. How are employees encouraged to express their ideas and concerns?

25. How does your company develop its employees?

26. Describe the corporate culture and personality.

27. What is the company's code of ethics and how is it communicated to employees?

28. Describe your company's management style and overall philosophy.

29. How does your company respond to changing market/economic conditions?

30. What are the company's plans for the future?

31. What are the company's goals, both short and long term?

32. What are the challenges your company is currently facing?

33. Describe a typical day.

Up to this point, I've mentioned many possible "good" questions that you can ask recruiters during an interview. Don't think, though, that ANY question you ask is good. You would be better off asking no questions at all rather than taking a chance on asking a bad (i.e. dumb) question. A bad question is one that could have been discovered through basic research, makes you look like your priorities are in the wrong place (i.e. too money motivated), or show a "What's in it for Me?" mentality. Most employers don't like students to ask questions about salary. As mentioned earlier, the topic of salary will come up if you survive far enough into the interview process. Let the employer initiate it. You should be well aware of a starting salary range for the position, so there shouldn't be a need to ask about it. Other areas, besides salary, that employers do NOT want you to ask about include:

1. The products or services the company offers
    - If you need to ask about these, why even bother to show up?

2. Asking about hours of work or overtime requirements
    - You're making it look like you're unwilling to put in extra effort.

3. Vacation time (don't ask about benefits, period)
    - You haven't even got the job yet, and you're planning your vacation?

4. Dress code
    - Might make the employer think you like to dress "different."

5. Anything that shows a "Me" mentality
    - How soon will I be promoted?
    - Will I have an office with a window?

6. How did I do? (Referring to an evaluation of how you fared in the interview)

It's a good idea to stay away from these areas. They make it look like you haven't done your research or that your priorities are a little "different" that what the employer would like. At this point you should be prepared, at least on paper, for your interview. Now it's time to figure out what to wear and how to succeed in the face-to-face job interview.

CHAPTER TEN – INTERVIEW ATTIRE

"Dressed to Kill…or at least get a Job!"

While your resume creates the employer's first impression of you the majority of the time, your appearance assumes the role of creating your first "in person" impression. Before the employer gets to even ask you one question, your appearance will have made an impact (for better or worse). Keep this in mind when you're pulling clothes out of your closet. The smartest thing to do when you're deciding what to wear is to stay conservative. You can take a chance and wear something a little "flashy," but you could find yourself inappropriately or underdressed when you arrive at the interview. That's not a chance worth taking.

On average, employers make the decision on whether or not to invite you back for another interview or offer you a job in the first four minutes of the interview. You don't want the employer to spend those first four minutes wondering where you shop or whether or not she should call the fashion police. There are guidelines based on your gender that you should follow when you're getting ready for an interview. If you're interviewing for a position in which fashion "sense" is a job requirement (i.e. fashion designer, buyer), you're allowed more flexibility with your choice of attire. Once you arrive at the interview, use the restroom (even if you don't have to go), because it's a good idea to re-check your appearance to make sure your tie is straight, hair is combed, and your nose is free of little buddies!

Here are some general guidelines:

WOMEN

1. Wear a business suit with a jacket (not too short or too tight). It's better to include a skirt, not pants with a suit. Stick with darker colors, as bright colors might make you stand out for the wrong reason.

2. Keep make-up and jewelry (No multiple rings or earrings) to a minimum. The more jewelry you wear, the less professional you look.

3. No high heels! Instead, basic low pumps will suffice. Make sure they're cleaned and polished.

4. Wear your hair conservatively.

5. Wear conservative nail polish (clear is preferred) if you wear any at all.

6. Minimal make-up. This isn't a beauty pageant.

7. Wear hosiery that is skin color or at least close. No rips or runs.

MEN

1. Follow the same conservative approach. Wear a dark (navy, black or dark gray) suit.

2. Wear a pastel or white shirt. Solid colors are the safest bet, but thin pinstripes are OK.

3. Wear a tie that is neither too wild nor flashy. Try a conservative pattern and avoid bright colors (silk is the best material to wear).

4. Wear conservative shoes that go with the suit and make sure they are clean and polished. Make sure your socks go up high enough on your legs so when you sit down you have no bare leg visible.

5. Shave before you leave for the interview and comb or brush your hair. If you prefer the beard and moustache look (although I would shave them off), make sure they are neat and trimmed. Getting a haircut is a good idea too if you have long hair.

6. If you normally wear an earring, avoid it in the interview because you're taking an unnecessary risk.

7. No rings on your fingers other than a wedding or school ring.

8. Cleaned and trimmed fingernails.

WOMEN AND MEN

1. Once you arrive at the interview setting, find a mirror and re-check all of the things you checked before you left your house or apartment (i.e. hair, clothing, etc.). You never know what might get out of place during the drive or walk to the interview.

2. Bring a professional looking pen to the interview. No cheap plastic pens, and do not bring a pen with a clicker at the end, as you will most likely click away and end up annoying the employer.

3. Bring a dark leather folder or light briefcase with extra copies of your resume and references. It is also a good idea to prepare some typed questions you plan to ask the recruiter. You do not want to have to rely on your memory when it comes time to ask questions.

4. Wear a nice watch. While a Rolex is not necessary, stay away from cheap looking plastic watches.

5. Make sure you are pleasing to the employer's nose. Do not overpower him or her with perfume or cologne, but make sure you do not smell like the inside of a gym either.

6. Make sure your breath smells fresh. Try a breath mint before you go into the interview.

7. Make sure your hands are clean and your fingernails are manicured.

8. No visible body piercing, other than ear piercing for women.

9. No gum, candy or other objects in your mouth.

10. Don't smoke. If you just finished a puff session, see #6.

11. Keep all potential noisemakers out of your pockets. Keys or change can be tempting to jingle if you have your hands in your pockets. Keep your pockets empty (hands included).

The general rule of thumb is to dress in formal business attire unless the company specifically tells you that it is acceptable to do otherwise. It's much better to show up to an interview in a business suit and find the employer in casual clothes than for the reverse to occur. Now that you're all dressed up with a very important place to go, it's time to cover taking part in the face-to-face job interview.

SECTION III

TAKING PART IN THE JOB INTERVIEW

CHAPTER ELEVEN – THE INTERVIEW OPENING

"Start with a Bang, not a Whimper!"

Returning to the commercial where you "only have one chance to make a first impression," chances are you won't get to make a second impression in the face-to-face interview if your first impression (resume and cover letter) was weak. If you do get over that first hurdle and make it to the face-to-face interview, you'll get a second chance to make a good impression when the employer greets you in the opening. The moment the employer lays eyes on you, an impression is made. This should emphasize the importance of arriving at the interview a few minutes early (15 as suggested earlier).

Employers are looking for several key things when assessing your first impression, and you'd be smart to try and show every one of them. First, smile! It seems smiling would be so easy. After all, you do it all the time without even thinking about it. So what makes it so hard to do in the opening of in interview? The answer is simple: NERVES! When people get nervous, they tend to be less comfortable and when they are uncomfortable they're going to be much less likely to smile. The problem with not smiling is that employers will see that non-smiling face and assume you aren't excited about the interview, the company, position, etc. They don't know how you're feeling on the inside. You have to show them how you are feeling on the outside. Believe it or not you're making it easier on yourself by smiling because it takes more muscles in your face to frown than it does to smile. It's not as easy as wearing a sign around your neck that says, "I'm really happy." Smile!

A second essential thing for you to do in the opening is make direct eye contact with the employer. When you're nervous it's easy to look away. The problem is that if you look away from the employer in the opening, she may conclude that you're lacking in self-confidence, which is a major problem. Because of the increased importance placed on the first impression, it's more important to make direct eye contact with the employer in the opening than at any other time during the interview. Of course, that doesn't mean that after you make a good first impression that you can stare at your feet the rest of the time. It also doesn't mean that you have to get into a staring contest with the employer. It's normally harder to maintain eye contact while you're talking. Unfortunately, you should spend roughly 3/4 of the interview talking, so if you're not making eye contact, the employer's impression of you is being negatively effected, and you can't afford this, especially in the opening.

A third vital part of a good first impression is to shake the employer's hand firmly. Nobody likes a wimpy handshake, especially employers. Make sure you clean and dry your hands, too, in the restroom before the interview starts. Make sure you dry your hands well because, more than likely, they're going to be a little sweaty because you're nervous. Wipe them off in the bathroom several minutes before the interview, not as you're standing face to face with her. Grab the employer's entire hand, not just the fingers. "Dainty" isn't a good word to use to describe an appropriate handshake. Once you have the employer's hand, shake firmly once or twice and then let go. You're not taking it with you so shaking it any more than twice can be a problem and can show your nervousness. What do you do if the employer doesn't extend her hand? You better extend yours anyway, in hopes that she'll extend hers. Even if you saw her leave the restroom without washing her hands, don't get caught with your hand at your side. She might be testing you to see if you'll take the initiative by offering your hand. Don't vary the firmness of your handshake regardless of the gender or size of the employer either. It doesn't matter whether the employer is a 250-pound male or a 100-pound female, give the same firm handshake. The last thing you'd want to do is offend a petite female employer by softening your normally firm handshake. If she concludes you did this because she's female, it could negatively effect her impression of you.

A fourth key opening element is being enthusiastic. That is probably very similar to smiling in that it's something you normally do all the time outside of a job interview. If you think about your day-to-day interaction with friends, family, co-workers, etc., I bet you envision an enthusiastic you. So why do so many students project a lack of enthusiasm in interviews? Blame it on those nerves again. Nervousness has the ability to turn enthusiasm into BORING! Don't let it happen to you. Make sure you TALK in the opening. Don't clam up. If the employer asks you how your day is going, don't just say "Fine." Go into detail. This is a great opportunity to impress the employer with your wonderful communication skills and outgoing personality. If you're missing those qualities, you better find them for your interviews.

As I mentioned earlier, by the time you get a face-to-face interview, it's not about your resume anymore. It's about your personality. Employers want to assess whether or not you would be an enjoyable person to work with. I know I want to work with personable and enthusiastic people and employers are no different, but you have to show them! It might help to think of the employer as a friend and you're just having a conversation. Try not to concentrate on the fact that this stranger may control your future. I hope you have an employer that takes time at the beginning of the interview to establish rapport and build some trust and warmth between the two of you, but that's certainly not a guarantee. If you interview enough you're bound to encounter some employers that aren't very friendly and make being enthusiastic difficult. However, it doesn't do you any good to match that attitude. While it might feel good to return that bad attitude, it won't help your chances of getting a job. It's not good enough to just tell an employer you're excited or enthusiastic about her company and the job. You have to show them enthusiasm.

As discussed in chapter ten, looking good in an interview is important. It is emphasized in the opening, because it plays a huge role in making your first impression. Looking good is supposed to include your professional appearance, not your attractiveness. Unfortunately, it includes both. If you're thinking to yourself, "I thought physical appearance was illegal to use as a qualification in the job search?" you'd be right. However, the reality is that it is used and it does make a difference. Everybody has an equal opportunity to show a professional appearance, which includes looking neat, being well-groomed, dressing professionally, etc. I would always wear professional attire (re-read chapter ten if you need to) unless you know for a fact that you can dress less formally. When looking at attractiveness, however, not everybody is equal. Attractive people have an advantage in the job interview. While that may seem unfair, blame it on the employers being human. People value good looks. They always have and they always will. Is it fair that a good-looking person gets a job offer instead of an equally or even more qualified, "less" attractive person? NO! Does it happen? YES! Don't let me scare you into thinking it's always that way, because it's not. There are plenty of employers that have the ability to hire the most qualified applicant regardless of attractiveness. There are other employers that assume attractive people have an array of valued characteristics. It may be that those good looks are clouding the employer's judgment into thinking a good-looking applicant is also more intelligent, motivated, enthusiastic, personable, etc. than less attractive applicants, even when they really aren't. If you want that unfair advantage in the job interview, I hope you're good looking.

One of the most important things to remember about the suggestions for making a good impression in the opening of a job interview is that they're easier said than done. You need to make a conscious effort to ensure that you're following them. Chances are very good that you're going to be nervous at the beginning of an interview. In fact, I'd have to check your pulse or administer a lie detector test if you said you weren't. I'm not just referring to your first interview either. You'll experience nervousness or anxiety, to some degree, before every interview because you never know exactly what to expect. This nervousness makes forgetting very easy and remembering what you are supposed to do and say very difficult. If you're the type of person who sweats bullets and is overcome by nausea just thinking about a job interview, then you need to ask yourself (not out loud) whether or not you're following these suggestions in the opening of your interview. If you don't even give them a thought, then you're probably going to suffer some negative consequences.

Most employers will try and make you feel relaxed and comfortable in the opening. They normally include some small talk over a subject that's easy for you to talk about, but unrelated to the job and company (i.e. the last movie you saw, how your classes are going, sports-related chit chat, etc.). While you can hope you have an employer that's out to make things as easy as possible for you, it's a wise idea to be prepared for anything, including the worst. Remember, try not to let it get to you if you have an employer that seems to be out to ruin your day. Stay friendly and focused. After the opening in an interview, you want to continue to shine by providing great answers to the employer's questions about you, the company and position, etc. Chapter twelve covers the qualities employer are looking and listening for when you're answering questions.

CHAPTER TWELVE - ANSWERING QUESTIONS

"Talk a Good Game!"

I referred to the job interview as a game in an earlier chapter. That is never more obvious than when an applicant starts answering questions. So many applicants just tell the employer what they think she wants to hear. The more you prepare for questions, the better you'll be able to do this. No doubt about it! There are really *two* parts in every answer to a question, a verbal and a nonverbal. The verbal part is the actual words that come out of your mouth. The nonverbal part is *how* you say them. They're both important, but many experts would argue that the nonverbal message is even more important and carries more of the message than the verbal. A key reason is that the nonverbal component is less controllable than the verbal. It's very easy to stare at an employer and tell her a bunch of make-believe information, but did you know while you're doing this, you're probably avoiding eye contact? You may not even be aware you're avoiding eye contact, but a skilled employer will. Regardless of which aspect of your answer you feel is the *most* important, both the content (verbal) and delivery (nonverbal) of your response are a crucial element in wowing the employer. Chapter thirteen covers the important nonverbal qualities of an answer. This chapter covers key verbal qualities that employers want you to include when answering their questions.

One key verbal quality entails being succinct. You might be thinking that means to reply with a brief answer. Nothing could be further from the truth. Being succinct means that you're "to the point" and only answer the question you're asked and don't ramble off onto unrelated topics. When you're asked an interview question, think about what it is the employer wants to know and provide only that information. The employer may want to know a lot. For example, that "Tell me about yourself" question is asking for a lot of information. Think about it. The only restriction placed on you with that question is that your answer has to be about you. However, you've probably been on this planet for at least 19-20 years, so you should be able to go on and on and on. The more information you can give the employer, the better. That requires less probing from the employer (i.e. less work for her, which is a good thing). Don't drift off on tangents during your response by talking about something unrelated to what the employer asked for. Feel free to give examples to support your answer even if you weren't asked. For instance, imagine you're asked, "What is your greatest strength?" Based on the wording, all that question is asking for is a brief response like "My leadership skills." It'd be a very good idea on your part to add a specific example of a time in your past you demonstrated your leadership skills, too (a popular example of a behavioral question from chapter seven). That wouldn't be considered going off on a tangent, because you are providing relevant information that will help illustrate and support your answer.

Another key verbal quality is making sure your answers are thorough and detailed. You might think that being succinct and being thorough and detailed are contradictory, but they really do go together like studying and getting good grades. You should strive for both. Answer the specific question you are asked but provide as much detail in your answer as you can to support yourself and your qualifications. As I already mentioned, this makes the employer's life easier (you're supposed to be the one breaking a sweat, not her) by allowing her to develop fewer secondary questions to obtain information she was hoping to receive from your answer to the primary question.

A third key verbal quality is providing well thought out and organized answers. The only way you can achieve this is by preparing for questions prior to your interview. I'd prepare for as many different questions as you possibly can before every interview. Look at as many different example questions as you can and think about what you would say in response to those questions. I didn't say *memorize*. Trying to memorize an answer is a very dangerous tactic on your part. If you do try, you might forget what you wanted to say during the interview and become very flustered. You also run the risk of the employer getting the impression that you're giving rehearsed responses and that could really undermine that sincere and honest image you're trying to project. The best thing you can do when reviewing possible job interview questions before an interview is come up with areas that you want to cover if asked a particular question. That way you're developing an outline of a response and not the actual verbatim response. For example, if you are asked the dreaded question "Tell me about yourself," prepare qualities you want to include like "I am a leader, I am motivated, I possess strong oral and written communication skills, etc." Then, rely on your ability to go into depth about examples of those qualities during the interview, but don't try to memorize examples in advance. That's too much to remember.

Chapter eight mentioned it's a good idea for you to understand going into an interview that there's no way you could ever prepare for every possible question that you might be asked. That'll make you less frustrated when you're asked a question you didn't prepare for and give you a better chance of calmly coming up with a response. One positive thing you'll discover as you take part in more and more interviews is that many of the questions employers ask will be the same from one interview to the next. You should find it a lot easier to answer a question you have been asked in a previous interview than one you have never been asked before. You have probably heard the expression "Fool me once, shame on you. Fool me twice, shame on me." That applies to interview questions. You're bound to be stumped a few times. Make sure though that each time it happens, you remember the question that gave you trouble and prepare so it doesn't happen again. If it does, you only have yourself to blame.

A final key verbal quality is making sure you appear sincere. You need to be honest with the employer and don't just tell her what you think she wants to hear. I know it's tempting, but as mentioned earlier, the more lies you tell, the more you have to remember. Many employers have mentioned to me they often have a difficult time in determining whether or not students are telling them the truth. If you start to answer a question immediately after it's asked, an employer might think you're giving a rehearsed response. You're better off pausing for a second or two, which also helps you collect your thoughts and helps you ensure you're answering the question you were asked. Not that you would actually be dumb enough to do this, but NEVER tell an employer you don't have any weaknesses (that in itself would be a weakness). She won't believe you. It's actually perfectly OK to admit a weakness if asked for one (don't volunteer them if you are not asked) and tell the employer how you're working on that weakness to turn it into a strength.

If you're hoping to find a section in this chapter that provides you with actual answers to possible employer questions, you're out of luck. Books that do provide you with the answers aren't really helping you at all. They're out there, but avoid the temptation. The answers need to come from you and represent who you are, not the author of some book. They need to be based on your own personal experience. Employers are hiring you, not the author. If you don't believe in yourself enough to provide your own answers, then why should an employer believe in you enough to offer you a job?

As an applicant in an interview, it's important to remember that you have rights. Those rights include not being asked questions that discriminate on an area that is unlawful to use as part of a hiring decision. In order for questions to be legal they need to inquire about an essential requirement for performing a job. There are numerous laws and statutes that have been passed by congress to protect you against discrimination. You can't be judged based on your race, sex, religion, national origin, ethnicity, marital status, disabilities, etc. If you're asked an unlawful question in an interview, what do you do? You are in a potentially no-win situation. If you choose to answer the unlawful question, you might not get the job because of your answer. If you decide to refuse to answer the question, you might not get the job because you appear uncooperative or difficult. As long as the question is not a severe infringement of your rights, there are several ways you can go about answering the question and be tactful at the same time:[2]

1. *Tactfully refuse* to answer the question
Recruiter: Do you plan on getting married?
Applicant: My marital plans will not interfere with my ability to perform the necessary duties and responsibilities of the position.

2. *Answer directly but briefly*
Recruiter: What is your religious affiliation?
Applicant: Catholic.
3. *Pose a tactful inquiry*

Recruiter: Are you married?
Applicant: Why do you ask?

---

[2] Gerald L. Wilson and H. Lloyd Goodall, Jr., *Interviewing in Context* (New York: McGraw-Hill, 1991), 159-162; Joann Keyton and Jeffrey Springston, "I Don't Want to Answer That! A Response Strategy Model for Potentially Discriminatory Questions," Unpublished Paper, Annual Convention of the Speech Communication Association, San Francisco, 1989.

4. Try to *neutralize* the recruiter's apparent concern

Recruiter: Do you plan on having children?
Applicant: Yes, I do. My husband and I are looking forward to the challenges of being parents and having a career. We have several friends who have met the challenge without any problems.

5. Try to *take advantage* of the question to support your candidacy
Recruiter: How would you feel supervising people who are older than you?
Applicant: I would not have any problem with this. I have worked with people older than me before and as long as I respect their needs, they respond well to my leadership, regardless of my age.

Remember that if you do get the job, you may be able to do something about the company's hiring practices. If the question is a severe violation of your rights as an applicant, you can refuse to answer the question and report the employer to his or her superior. I highly doubt the company is aware that she is asking unlawful questions and once they were informed, they would most likely stop this behavior. Employers know that it can cost them hundreds of thousands of dollars if they lose a discrimination lawsuit. Even if they're sued but win in court, it can still do irreparable damage to the company from a public relations standpoint.

CHAPTER THIRTEEN – NONVERBAL BEHAVIOR

"Actions Speak Louder than Words!"

It's true, most employers do think actions speak louder than words. You can say all the right things, but if your actions contradict what you're saying, this can lead to a negative perception of you by the employer. Most people tend to believe your nonverbals when they're conveying a different meaning than your verbals. There are numerous nonverbal behaviors you need to *pay attention to* during an interview in order to ensure they're consistent with the words you are speaking. "Pay attention to" is italicized because if you don't, chances are good you'll revert back to old form. That means if they were a problem in the past they'll be a problem again and you don't want them resurfacing in an interview.

How do you know if you have a problem with one or more nonverbal qualities? Think about the last time that you delivered a speech or took part in a group presentation. How were you evaluated on your nonverbals? If you received two thumbs up, you should be okay. If you didn't, those same problems will probably sneak back into your repertoire of behaviors during an interview. That's something you want to avoid. The best way to avoid those unwanted nonverbals is to "check in" with them occasionally as you're taking part in an interview. For example, imagine that on your last speech your teacher told you your eye contact was poor. You looked down at the ground too often. You might think that you don't have to worry about this in an interview. After all you did it in a speech, not an interview.

An interview and a speech are the same in many ways. You're nervous in both. You're being evaluated in both. Both put you "on the spot" and make you do things when talking that you wouldn't normally do. That means if your eye contact was poor in a speech, it'll probably be poor in an interview. The best thing to do, as mentioned earlier, is to "check in" with any nonverbal behaviors that you've had a bad experience with in the past. Ask yourself (not out loud) if you're making direct eye contact, particularly when you're answering a question. If you notice you're looking away, make sure you focus back on the employer. If you don't give your nonverbals a second thought during an interview, chances are good the employer won't give you a second chance. There are several nonverbal qualities that you need to show an employer in an interview so she thinks your actions are repeating the same message as your words.

First, maintain good eye contact. The importance of eye contact has already been discussed several times. Just remember that if it's lacking the employer might assume you don't have a lot of self- confidence or that you are being dishonest or something else that will negatively effect her perception of you. Don't give the employer the chance! Keep track of your eye contact to make sure it is focused on the eyes of the employer, especially when you're answering a question.

Two nonverbal qualities that feed off of one another and are both very important are sounding enthusiastic and speaking in a conversational style. If you're enthusiastic about what you are saying to the employer, then you're most likely speaking in a conversational style. If you're unsure of what a "conversational style" is, think of a professor you had in college that would best be characterized as "monotone." His (I'm using "his" because it is much more common in men) tone didn't fluctuate at all when delivering a lecture. Did that professor *sound* excited to you? Was it enjoyable to listen to his lecture? I'm sure your answers to both are an emphatic "NO!" The employer won't be any more interested in listening to you than you were in listening to your professor. You need to portray yourself as a dynamic individual with lots of energy. One important aspect you're evaluated on in an interview is the employer's perception of how well you'd "fit" in the organization. This doesn't mean how physically fit you are. It deals with your "interpersonal fit." Will you be a person that other members of the organization will enjoy working with? That's a huge factor in the face-to-face interview. Your paper qualifications (i.e. resume, cover letter, etc.) can speak for themselves, but once you get the interview, it is all about your personality. Employers hire personalities, not resumes. Be enthusiastic!

Another important nonverbal quality is good body language. Hopefully, your body isn't making ANY noise during your interview. Good body language doesn't have anything to do with that anyway. It involves several things. One is your posture. You need to sit with a good posture. The best posture in an interview is usually not the most comfortable. Sit up straight in your chair and lean slightly forward (not so much so that it looks like you're going to tip over) as that is an indication of your interest. Keep your legs together. You can cross them at the ankles if that is comfortable for you. If you're female, you can cross them at thigh level (I wouldn't cross at the thigh if you're male even if you can), but be careful not to shake that top leg. Males have the tendency to "spread out" when they sit by putting their feet on the chair in front of them or slouching in their chair. Guys, don't take up as much space as you normally do. Keep your hands in your lap, holding a professional leather folder (or something similar).

A second part of body language involves your gestures. If you like to add a lot of hand movement when you talk, be careful not to add so much that your hands are flying in front of your face. Be aware that your body is going to produce an extra amount of adrenaline in an interview because it likes to do that when you are nervous. That adrenaline is going to result in "bottled up" energy that is dying to come out. Don't let it kill your chances at a job. Channel that energy towards positive nonverbals (increased volume, more projected enthusiasm, etc.) not negative (i.e. shaking your leg, tapping your feet on the floor, fidgeting in your chair, excessive hand gestures, playing with your pen, etc.). Don't bring a pen that has one of those noise-making clickers at the end as you're bound to start clicking it without even knowing it. Good gestures are ones that, for the most part, go unnoticed by the employer.

Speaking at an appropriate volume level is another important nonverbal quality. If you're naturally a soft-spoken person, make an effort to speak up during your interview. A low volume can also project a lack of self-confidence. Speaking too loud is not normally a problem, as people usually get quieter when they're nervous. However, if you are a person who has been described as boisterous (a nice way of saying you are loud), remember the employer is only a few feet away and there's no need to shout. If you're going to error on one side or the other, I'd error on the side of being too loud. An increased volume level will normally be rated as positive because you're projecting more energy and enthusiasm, which indicates you're more interested in the interview.

Speaking at an appropriate rate is also important. When people get nervous they usually speak at a faster rate, sometimes so fast they find themselves having trouble enunciating their words. It's actually a good thing to speak at a slightly accelerated rate. The employer will evaluate you more favorably as a slightly accelerated rate projects more energy and enthusiasm. Don't get carried away, though. Slow down if you're speaking so fast that your brain can't keep up with your mouth.

Confidence is a quality that has both verbal and nonverbal parts to it. You can say specific things in your response to show your confidence in your skills and abilities like "I work very well in teams" or "I have very strong leadership skills." You can also sound confident by projecting many of the important nonverbal qualities (eye contact, enthusiasm, volume, etc.). Confidence is a package deal. You'll need to have both the verbal and nonverbal qualities. They'll work together to project a lot of or a lack of confidence. The employer can look at you and listen to you for just the first minute or two in the opening (another indication of the importance of a strong first impression) and tell whether or not you're confident. Try and include as many verbal and nonverbal qualities as possible, because confidence is one quality that every employer is looking for.

If your actions are going to speak louder than your words in an interview, then you need to make sure they are saying the right things. Since many nonverbals occur without you even knowing it, remember to check in with them occasionally during the interview. If you follow this suggestion you'll be able to correct them if they're working against you. If you don't follow this suggestion, you run the risk of tapping a tune with your pen the employer doesn't enjoy.

CHAPTER FOURTEEN – THE INTERVIEW CLOSING

"Finish with a Flourish!"

Chances are good you'll know when the employer is in the process of closing the interview. A verbal closing ("That's all the time we have today. Thank you for coming in...) or a nonverbal closing (employer closes a notebook, leans forward in her chair, stands up, etc.) can be used to signal the end of the interview. Now isn't the time to let your guard down by saying or doing something stupid that might ruin an otherwise solid interview. You worked hard (or at least you should have) to prepare for the interview. Don't let all of that preparation go for naught. Some employers operate under a "recency effect" which means they'll remember most what they read, see or hear last. That means your closing may be the most memorable part of the interview for them so put as much effort into your closing as you did in your opening. Most employers say that if you haven't made a good impression on them by the closing, there's nothing you can do to breathe life back into your chances of getting another interview or a job offer.

One thing you can do to ensure a strong finish is reiterate your interest in the position and company. If you wait until the closing to show your interest in the company and position, you're a little too late. You need to let the employer know in the opening and in the body of the interview that her company is clearly the one you want to work for. Tell her you're excited about the prospects of working there and her company is number one on your list (you hope you're number one on the company's list, too). None of these comments guarantees you anything, but they certainly won't hurt your chances. An excellent way of indicating your interest in the body of the interview (as discussed in chapter six) is to show the employer how much you researched by giving information on the company and position or asking questions that clearly demonstrate you've done your research. Don't get carried away and tell the employer every five minutes that her company is number one as that might be perceived as going overboard and that you're just kissing up. Ideally, you want to be able to demonstrate your interest in the company and position without having to verbalize it.

The second suggestion, restate why you would be an asset to the company, also indicates that you're doing something "again" in the closing. To be able to accomplish this you must have already convinced the employer that your skills, abilities, knowledge, training, experience, etc. make you the best person for the job. You probably won't have a lot of time to reiterate why you would be an asset in the closing. You might even feel a little uncomfortable doing this as it could make you think you are laying it on a little thick. That's why it is so important to accomplish this during the body of the interview when you have more time, so you do not have to rely on attempting it in the closing. However, it doesn't hurt (if it seems like you have some time in the closing) to mention one or two of your strongest selling points. These selling points ideally would be qualities you possess that not every other student would. Of course, if you did not prove to the recruiter during the body of the interview that you would be an asset to the company, attempting to do this in the closing might very well be a futile effort.

Make sure you thank the employer. Sure she's getting paid for it, but you still want to express your appreciation. Employers like to hear your appreciation even though it's their job to conduct the interview. It demonstrates respect, consideration and courtesy on your part. Thank her for showing interest in you as an applicant and taking the time to conduct the interview. Tell her how much you enjoyed meeting and getting to know her (even if you really didn't). It's also a good idea to say that you're looking forward to (hopefully) becoming a member of their organization. Basically, you want every employer to think they're number one on your list. Sure, they can't all be number one, but they don't have to know that. Or you could choose to be honest and tell the employer, "Well, you're number five on my list, and if none of the top four offer me a job, I'll be ready to go to work for you." Okay, Okay. I've preached about being honest up to this point. Why am I changing my mind? I'm really not. This is just one of those times that a little white lie doesn't hurt anyone, so I'd recommend it.

Several key aspects of the opening are very important in the closing, too. Give the employer a good, firm handshake and make good eye contact throughout the entire closing (don't shake her hand the entire closing, just make eye contact). You should clearly know by now how important those two suggestions are. They may be a little easier for you to do in the closing because by this time you should feel pretty comfortable with the employer. What if you were one of those unlucky students that had a "bad" employer that actually made things less comfortable by the end of the interview? Or maybe you had an employer conduct a stress interview? It doesn't matter. Even though you may feel like making a fist and making some contact with the employer's eye, it's important not to let your frustration get the best of you. Look the employer in the eyes, shake her hand firmly, and try to smile.

Every applicant wants the employer to explain exactly what to expect after the interview. This doesn't always happen. You may get an employer that tells you in the closing "I'll be in touch soon to let you know whether or not you will be extended an offer." When is "soon?" Tomorrow? Two weeks from now? The problem is you really don't know. What if you thought "soon" meant tomorrow and at the end of tomorrow you still had not heard back from her? You try calling the company because the employer must have just lost your phone number. Why else would she not call you back? However, in the employer's mind when she said "soon" at the end of the interview, she meant two weeks. In this scenario you'd probably be perceived as being overanxious or too aggressive. On the other hand, if you thought soon meant two weeks but the employer thought two days, you could lose a job because you didn't follow up in time. It's your right to know exactly when you can expect to hear back from the company. Ask the employer when you'll hear from her if it's not clearly and explicitly communicated to you in the closing. Knowing when you will hear back from the company will also help you if you have a deadline for accepting a job offer that has been extended to you, while at the same time you're waiting to hear back from the number one company on your list.

There's a very funny episode of Friends where Chandler's taking part in a job interview. The employer tells Chandler about his job "duties" and of course Chandler's childish mind thinks "doodies." He does a valiant effort of fighting back the laughter. In the closing of the interview, the employer tells Chandler he did great and that he can say with some certainty, the job is his. Chandler's thrilled and, feeling pretty comfortable now, lets the employer in on his secret by telling him he had a hard time not laughing at the doodie question. The man looks at Chandler in disbelief. Chandler proceeds to say the word "poo" to the employer (you never want to bring that up in an interview) when it's clear he doesn't understand. Chandler realizes he's blown it and tries to convince the employer the interview was already over and this didn't count. Too late!

The moral of Chandler's interview story is that it's a good idea to get it engrained in your head now that until you're out of sight and mind of the employer who conducted your interview, the interview is not technically over. Anything you say or do could still be used against you in the employer's evaluation. Pay very close attention to your comments and actions in the closing because there's that natural tendency to let your guard down (not that you would ever talk about poo) when you realize the interview is coming to a close.

Make sure you get a business card from the employer before you leave the interview setting so you can use it to write her a thank you letter. The last thing you want to do, especially if you had a really good interview, is spell the her name incorrectly. That'd be very embarrassing and potentially costly, in terms of a job. It's also a good idea to ask if there is anything else you can provide (i.e. references, transcripts, etc.) Once you're completely removed from the interview situation, it's time to begin the follow up process.

SECTION IV

FOLLOWING UP AFTER THE JOB INTERVIEW

CHAPTER FIFTEEN – THE FOLLOW UP

"The Day After!"

If you're like most students, you'll experience a huge sense of relief after you leave the interview (whether you did well or not). You need to realize that the job search process is not quite over. If you believe in and act on the expression "Good things come to those who wait," you'll find yourself waiting without many good things coming. A more appropriate suggestion for following up after an interview is "He or she who hesitates is lost." The day of or the day after an interview you need to take action. This action comes in the form of a thank you letter (Figure 15.1) to the employer that conducted your interview. Remember the importance of getting a business card so you know the name of the person you're sending the letter to. Personal, short and sweet are good adjectives to use when describing a thank you letter. It's also a very good idea to mention something specific from your interview with the employer (i.e. personal) that will trigger the memory of you and your interview in the her mind. Needless to say, make it a positive association.

Most employers feel a thank you letter is important. Ten years ago you could send a thank you letter and an employer would think, "Wow, Mary's such a conscientious and thoughtful person." This could only help Mary in her quest for a job. Unfortunately, now most everyone sends a thank you letter but that doesn't diminish the significance of sending one. Failing to send a thank you letter today will make you stand out in a not-so-positive way. In this case the employer might think, "Wow, I received a thank you letter from every applicant except Mary. I guess she had more important things to do or doesn't really care about me or the company." This could result in your credentials being filed in the circular file. This may be an extreme example but the message should be clear: You need to send a thank you letter just to keep up with the rest of the applicants who are sending one, so you don't unnecessarily hurt your chances of getting a job.

Use the employer's business card to make sure you get the correct spelling of her name. Not every name's as simple to spell as Mary Smith. One decision you have to make concerning the thank you letter is whether or not to type or write the letter by hand. Let the relationship you established with the employer during the interview dictate that decision. If your employer was very personable, made you feel very comfortable or encouraged you to refer to her on a first name basis, I'd hand write the letter. You can write the letter on business stationary or a business card. Crane's is a stationary company that makes very classy stationary and note cards, ideal for a thank you letter. If you didn't get a good feeling from the employer (i.e. no first names, very little rapport, etc.) I'd type the letter on business stationary. Always error on the side of being conservative. If you're having trouble deciding whether to type or handwrite the letter, typing it would be the safest suggestion. If you don't know the marital status of a female employer, address it to Ms. Smith, not Mrs. Smith.

As long as the employer told you when you could expect to hear from her concerning another interview or a job offer, you should know when it's okay for you to call to check on your job status. As mentioned in chapter fourteen, if no mention is made of a time frame, error on the side of being aggressive so you don't end up waiting by the phone for a call that never comes. For jobs that are more aggressive than typical jobs, like sales, it is a good idea to call back every couple days unless an employer specifically tells you not to. The employer may be assessing your aggressiveness and if you don't make the call, you might not get the job.

There's an example thank you letter in the appendix. It's an example of a cover letter written to an employer that conducted a more formal interview and not on a first name basis. It's hard when you are typing a book to include an example of a hand written thank you letter. The hand written letter differs from the example you see in that you would not have your address and the company address at the top of the letter. It would be on the envelope only. Also, you would have the letter addressed to "Dear Beth" and not "Dear Ms. Fordham." The body of the letter would stay roughly the same although you could add another comment or two on something personal that you and the employer discussed (i.e. "It is always neat to talk to another animal lover").

After the follow up process, it's time to start sifting through all those job offers that start coming your way. Chapter sixteen offers advice on what to do with them.

Figure 15.1 – **Sample Thank you letter**

123 Westwood Way
Tustin, CA 92117
January 22, 2006

Ms. Beth Fordham
Marketing Manager
Rivercrest Pharmaceuticals
634 Wooded Pines
Cincinnati, OH 45263

Dear Ms. Fordham:

Thank you for meeting with me today. I enjoyed our discussion on life as a Marketing Manager at Rivercrest Pharmaceuticals and I appreciate your insights regarding the shift in the way manufacturers are marketing products today.

Rivercrest Pharmaceuticals offers the opportunity to work for a top marketing organization that places a premium on innovative, creative and analytical thinking. This is exactly the type of environment I am seeking.

I am excited about the opportunity to work at Rivercrest Pharmaceuticals and appreciate you taking the time to talk with me. My skills and experience in making teams more cohesive and productive will make me an asset to your organization. I look forward to hearing from you and can be reached at (572) 645-3529.

Sincerely,

(Signature)

Courtney L. Davis

CHAPTER SIXTEEN – THE JOB OFFER (S)

"Show Me the Money!"

Hopefully, at the end of the interview process, you're in a situation where you're having to decide between different job offers rather than wondering why you haven't received any. Once those offers do start rolling in, you're going to have to figure out which one is the best for you. To accomplish this task, re-check the list of important characteristics and values (chapter three) you're looking for in a job. List the strengths of a particular job in one column and the weaknesses in a separate column. Creating a list is a good way to ensure that you are taking everything into consideration and not overlooking an important factor when deciding which job offer to accept. Some additional possible factors to consider include:

- Starting date
- Duties and responsibilities
- Opportunities for advancement
- Workplace environment
- Employees with whom you would be working
- Management style and philosophy
- Hours worked
- Insurance
- Amount of travel involved
- Job location
- Tuition reimbursement if you want to go to graduate school
- Moving expenses
- Flexibility
- Benefits (There are numerous possible benefits to consider)
    - Overtime (always look into whether OT is paid or not as it can make a big difference in your annual salary)
    - 401 (k) plans
    - Stock ownership
    - Flextime
    - Childcare
    - Employee assistance programs (EAP)
    - Retirement benefits
    - Salary and compensation

If you're interested in comparing salary offers from companies in different cities to see the effect of the cost of living, use the Salary Calculator (it compares the cost of living in hundreds of U.S. and international cities) at the following Internet address:

http://www.homefair.com/homefair/cmr/salcalc.html

This is how it works. Imagine you were offered a job in Pocatello, Idaho that paid you $25,000 annually. You are also interviewing for a position in San Francisco, California. To live at the same standard of living, your salary would need to be over $59,000. In other words, you shouldn't be jumping for joy if the employer in San Francisco offers you $30,000. You probably wouldn't even be able to make ends meet. As you can plainly see this would be a very good thing to check into if you are deciding between offers in different cities. This web site also assists with home and apartment relocation, calculating moving expenses, etc.

If you're like most people, you'll have certain qualities in the job you're seeking that are more important to you than others. For example, it may be of utmost importance to you that you have a high starting salary. At the same time, you're willing to work anywhere if it will take care of your need for the big bucks. Regardless of which job quality is most important to you and which is least important to you, making a list and checking it at least once is a good idea. Here's an example to consider:

SCENARIO:

You're offered your first job. On the positive side it offers excellent opportunities for advancement. The location is in Dallas and the company will cover your moving expenses, two factors that you wanted although they were not of major importance. Also, you can start in August, which is nice since you just graduated in May and wanted to take the summer off. You're happy with your duties and responsibilities and the work environment is one in which you would be able to work comfortably. The job has a great pension plan even though that is the last thing on your mind at this time.

On the negative side, it doesn't pay what you were hoping and it will not cover your expenses if you plan to continue your education and get an advanced degree, both of which were very important to you. It requires more travel than you would like and you really wanted better insurance coverage because your health record is not very good. You also learned that overtime is normal which does not appeal to you very much since You'd rather put in your eight hours and call it a day.

STRENGTHS	**WEAKNESSES**
Advancement Opportunities	Salary
Location	Tuition reimbursement
Moving expenses	Travel involved
Starting date	Insurance coverage
Duties/responsibilities	Hours worked
Work environment	
Pension plan	

The results are interesting and point out the value of taking the importance of each quality into consideration. The strengths of the job are greater in number, but when you factor in the importance of each job quality, the weaknesses would probably come out on top. The point is, don't automatically assume the job offer with more positive qualities is better because, as you can see from the list, if extremely important qualities are not satisfied that might count more in your overall decision.

Negotiating a Job Offer

Many people don't feel comfortable trying to negotiate with the employer and that might very well be due to a lack of experience in the process of negotiating. If you have an idea of how the process works, you'll probably be much more comfortable when the opportunity presents itself. The first thing you should do is determine exactly how much you'd require in your starting salary. Remember to let the employer initiate the topic, but you definitely need to know what you are worth. It's a good idea to research salary figures for positions and remember to take the cost of living into account.

Sources to use to research salaries:

- JobStar Salary Info   (jobstar.org/tools/salary/sal-prof.htm)
    - This is the best list of salary surveys on the Internet, in my opinion. It has over 200 salary surveys.
- Salary.com   (www.salary.com/home/layoutscripts/sall_display.asp)
    - Between the first two, you should be covered
- CareerBabes Salary Sites   (www.careerbabe.com/salarysites.html)
    - Includes salaries of men versus women
- American Almanac of Jobs and Salaries
- National Association of Colleges and Employers (NACE): Salary Survey
- People currently working in the field
- Professional association surveys

Once you've established your value, along with the value the industry has placed on a person in your position, you're ready to take part in the art of negotiating. The process begins when you receive an offer from the company. Determine where the salary falls within your desired range. Always remember to calculate the employer's entire compensation package (i.e. relocation expenses, tuition benefits, overtime, etc.) along with the salary. It's not recommended to jump at the employer's first offer, even if it makes you do cartwheels on the inside. You could be perceived as a little desperate in this scenario. If the salary is below what you were hoping for do not get discouraged. Often times the initial offer is intentionally low. If they can get you for less money they will. You need to know your options. Realize that you can always try to ask for an early review of your performance, which can lead to an early raise. You can ask for benefits besides salary including tuition expenses, stock options, etc. It's a good idea to be flexible in your demands.

One advantage you have when an offer is extended to you is that, no matter how good or bad it is, that's a clear sign that the employer likes you. That places a little more control on your side. In many ways the negotiation process is about who has more control, you or the company. On the surface it might appear as though the company has more control. That makes sense because there is only one person guaranteed to leave the interview with a job and that is the employer. There are ways, however, for you to increase your bargaining chips (i.e. control over the desired outcome, whether it's a job offer or a good salary and benefits package, etc.).

Bargaining Chips

1. Your qualifications – your education, work experience, skills, knowledge, abilities, training, etc. The more qualified you are in these areas, the more bargaining chips you'll have in the negotiating process.

2. Interviewing skills – the more skilled you are in the art of interviewing and communicating, the more bargaining chips you'll have. The key to many successful interviews and negotiation sessions is the personality of the applicant.

3. Economy/Job market – during strong economic times when a company is prospering economically, it may have more available jobs, which would increase your control. During good times you would also have offers from other companies to choose from, which would give you more bargaining chips. However, if jobs are scarce, this would probably have a detrimental effect on your control. The employer in this case knows you're probably having a more difficult time finding a job, which could result in a "low ball" offer (the bare minimum). It knows if you do not accept it, there'll be another desperate student coming down the path very soon.

Unfortunately, the employer has a crucial bargaining chip mentioned earlier that you don't have going into the interview process, a job. That in itself gives her more control over the interview process because she knows there's less on the line and less to lose. Oh sure, the company may lose the best possible employee ever (YOU), but regardless of the outcome for you, the employer will still leave with a job. If you have other job offers on the table, that'll give you more control because you won't need any one job as much because you have other offers.

If you feel like you have a lot of bargaining chips and are dissatisfied with a employer's original offer, try to negotiate for a better one. Be careful not to offend the her though. Be tactful and gracious in everything you say and be realistic in your expectations and requests. Remember that if you accept the employer's offer, you can work for a little while and demonstrate what you knew all along: You're the best employee the company has ever had. Once they realize this, the negotiation process becomes much easier.

Accepting or Declining a Job Offer

There will come a time (at least you hope there will) when you'll need to either accept (Figure 16.1) or decline (Figure 16.2) a job offer. Be grateful for the offer, especially with those you decline. As I've mentioned before, you never want to burn a bridge. You're probably going to be rejected by an employer or two in your job search. Most rejected applicants refer to the letter they receive in the mail as a "ding" letter. It's a wise idea for an employer to personalize and be sincere in a ding letter, but often times they don't. A guy friend of mine once received a rejection letter addressed to Mrs. Newlin. It certainly didn't make him feel like the employer went out of its way to be sincere and let him down easy. If employers make a habit out of this insincere tactic, it could ultimately end up in a lower number of quality applicants. Students talk to one another about their interviewing experiences and no one wants to work for an employer that doesn't seem to genuinely care about its employees. Take time out to write a nice rejection letter because you never know what the future holds and you could end up interviewing with that company again somewhere down the road.

Figure 16.1 – **Sample Acceptance Letter**

1425 Temple Avenue
Boston, MA 46358
June 15, 2006

Ms. Arline Montgomery
Director of Recruiting
Top Investment Firm
Chicago, IL 75645

Dear Arline:

This letter acknowledges my acceptance of the Investment Counselor position. I am very excited and enthusiastic about the opportunity to work with Top Investment Firm. I am confident I will be able to make a significant contribution and fit in admirably as an employee.

I understand the details of the offer as outlined in your letter. I will be arriving in Chicago on Friday, June 22nd and will be ready to begin work on Monday, June 25th.

I truly appreciate your confidence in my skills and abilities and look forward to working with you at Top Investment Firm.

Sincerely,

(Signature)

Drew M. Koltun

Figure 16.2 – Sample Rejection Letter

4635 Crystal Drive
Holly, CT 63524
May 20, 2006

Mr. Frank Dorazio
Regional Manager
Taffy Industries
Colorado Springs, CO 24075

Dear Frank:

I am writing to thank you for the generous job offer at Taffy Industries. However, I am going to have to decline the offer. As a student, I have to consider numerous aspects of the different offers I receive. After doing so, I have decided to narrow down my search.

I would like to thank you for everything you did to help me along the way. I also want to let you know that everyone who was involved in the interview process treated me professionally and with respect. Removing your company from consideration was a very difficult decision for me.

Thank you again for your expressed interest in me and for understanding the position I am in. I would not hesitate to recommend any of my friends to seek employment with Taffy Industries. I wish you and TI the very best of luck in the future.

Sincerely,

(Signature)

Steven P. Meyers

The next step in the process (starting your career) is up to you. This book is designed to help you get a job, not work it for you. Hopefully you'll find success in your job search. This book should help you on your way. If you follow the advice mentioned in these pages, your chances of getting job offers should increase. Keep it in your personal library even after you finish college as many students end up changing jobs a year or two after graduation and it can help you all over again. Your parents may know what's best for you most of the time, but when it comes to the job search process trust those people "in-the-know." Employers know best and their answers are represented here. You need to realize that being rejected for some jobs is perfectly normal and happens to just about every student. Try not to take rejection personally and let it get you down. That might end up affecting your self-confidence in future interviews.

There you have it. Everything you need to know to get a job. Now you just have to go get one. My hope for all of you is that you find a job that you love. Money's nice, but to be able to get up out of bed in the morning and actually look forward to going to work… that's awesome! I love what I do and I hope you all can say the same, very soon! I wish you the best of luck in your search for a job that you love!